VOLCANOES AND THE EARTH'S INTERIOR

Readings from
**SCIENTIFIC
AMERICAN**

VOLCANOES AND
THE EARTH'S INTERIOR

With Introductions by
Robert and Barbara Decker
Dartmouth College

W. H. Freeman and Company
San Francisco

All of the SCIENTIFIC AMERICAN articles in *Volcanoes and the Earth's Interior* are available as separate Offprints. For a complete list of articles now available as Offprints, write to W. H. Freeman and Company, 660 Market Street, San Francisco, California 94104.

Library of Congress Cataloging in Publication Data

Main entry under title:

Volcanoes and the earth's interior

  Bibliography: p.
  Includes index.
  Contents: The subduction of the lithosphere / Toksöz—The crest of the East Pacific Rise / Macdonald and Luyendyk—Hot spots on the earth's surface / Burke and Wilson—[etc.]
  1. Volcanoes—Addresses, essays, lectures. 2. Earth—Internal structure—Addresses, essays, lectures. I. Decker, Robert Wayne, 1927–   . II. Decker, Barbara, 1929–   . III. Scientific American.
  QE522.V8984   551.2′2   81–15092
  ISBN 0–7167–1383–7   AACR2
  ISBN 0–7167–1384–5 (pbk.)

Printed in the United States of America

1 2 3 4 5 6 7 8 9 0   FL   0 8 9 8 7 6 5 4 3 2

PREFACE

Volcanoes have their roots deep in the Earth's mantle and scatter their ashes high in the stratosphere. Because of this diversity there is no single science of volcanology; instead, volcanoes are natural laboratories where scholars of many disciplines study together. Geologists, physicists, chemists, meteorologists, mathematicians, biologists, archeologists, historians, artists, and philosophers have all contributed to our present understanding of volcanoes.

Abraham Werner, an eighteenth century geologist, dismissed volcanoes as freak furnaces where burning seams of coal melted sedimentary rock that had formed in a universal ocean. However, James Hutton, a contemporary of Werner, saw volcanic rocks as primary products from the Earth's interior. Nineteenth century geologists, agreeing with Hutton, assumed that the Earth had a thin, solid crust overlying a molten mantle. Volcanoes were easy for them to explain: just crack the crust. By the turn of the twentieth century, seismologists had determined the mantle to be solid, as strong as high-alloy steel; again, volcanoes became harder to explain. After mid-century, seismologists, probing in greater detail, reappraised their data, allowing for a partially molten layer in the upper mantle.

As the concept of moving horizontal plates of the Earth's crust was being reexamined in the 1960s, J. Tuzo Wilson, geophysicist at the University of Toronto, noted that volcanoes in oceanic chains like Hawaii are progressively older in the direction of plate motion. His idea was that some volcanoes form over a mantle hot spot and are then carried away with a slowly moving crustal plate, making room for a new volcanic island. This added a key argument to the concept of sea-floor spreading.

Meanwhile, geochemists analyzing volcanic rocks from many settings have determined that the mantle source material must be quite heterogeneous. Some source regions appear to be more completely or more repeatedly melted than other regions, suggesting complex and differing evolutions for various parts of the Earth's interior.

Other contemporary concerns about volcanic activity are voiced by scientists other than geologists. Meteorologists debate whether—and, if so, to what extent—volcanic dust affects weather and climate. Statisticians study the eruptive patterns of volcanoes to try to forecast their behavior, and botanists and zoologists examine the development of plant and animal ecologies on newly formed volcanic islands and on landscapes sterilized by lavas and pyroclastic flows.

Humanists as well as scientists become involved with volcanoes. From the volcanic burial in 1500 B.C. of Akrotiri, a Minoan city on the Aegean island of Thera, archeologists, historians, and geologists are piecing together a possible answer to the origin of the Atlantis legend. Even artists find volcanoes hard to resist; the famous Mexican painter Dr. Atl bought Parícutin Volcano from the farmer in whose field it was erupting so he could own as well as paint his passion. In various religions and philosophies, volcanoes are the vents of hell, the forges of weapons, the chaos of the underworld, or the mischief of a beautiful but tempestuous goddess.

The scientific study of volcanoes has grown rapidly in the past decade, yet in the words of a Los Alamos physicist it remains "primitive." But if volcanology is primitive, it is also wonderfully complex. That is what makes studying volcanoes more than just an occupation; as it was for Dr. Atl, for most of us it is a passion.

This collection of articles from SCIENTIFIC AMERICAN reflects such enthusiasm and also provides an excellent sampling of recent studies in volcanology. Part I explores the regional settings of volcanoes with regard to plate tectonics; Part II deals mainly with the differing products of explosive and nonexplosive volcanoes; and Part III shows volcanoes as dim windows on the products, processes, conditions, and evolution of the Earth's interior.

Will Durant noted that "Civilization exists by geological consent, subject to change without notice." It behooves us all to know more about our restless Earth.

June 1981   Robert and Barbara Decker

CONTENTS

Notes on cross-references to SCIENTIFIC AMERICAN *articles*: Articles included in this book are referred to by title and page number; articles not included in this book but available as Offprints are referred to by title and Offprint number; articles not included in this book and not available as Offprints are referred to by title and date of publication.

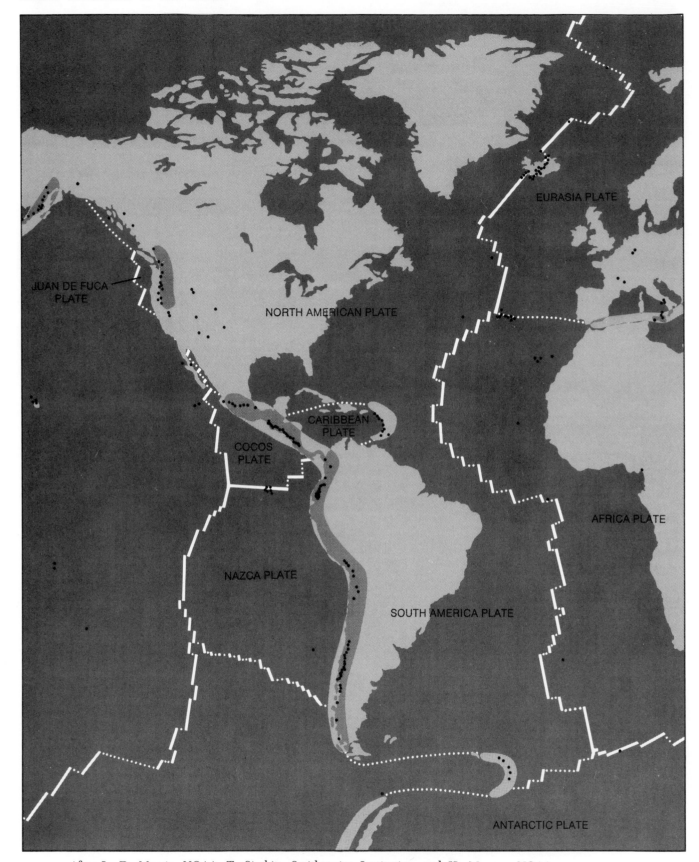

After L. D. Morris, NOAA; T. Simkin, Smithsonian Institution; and H. Meyers, NOAA.

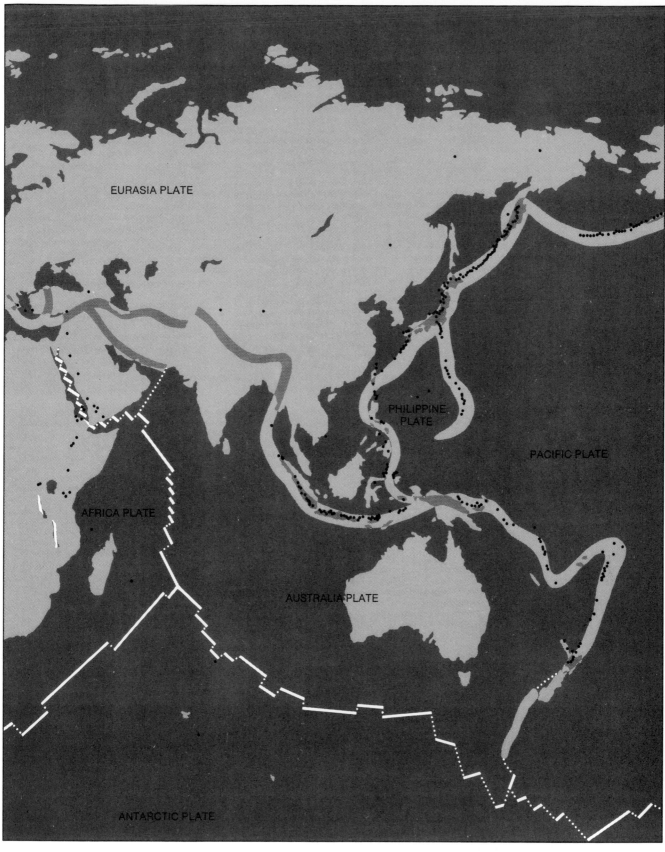

EURASIA PLATE

PHILIPPINE
PLATE

PACIFIC PLATE

AFRICA PLATE

AUSTRALIA PLATE

ANTARCTIC PLATE

KEY

 RIFT ZONES

 STRIKE-SLIP (TRANSFORM) FAULTS

SUBDUCTION ZONES

• GEOLOGICAL YOUNG VOLCANOES

VOLCANOES AND THE EARTH'S INTERIOR

VOLCANOES AND
PLATE TECTONICS

VOLCANOES AND PLATE TECTONICS

I

INTRODUCTION

An active volcano is one that has erupted during recorded history. This definition is as uneven as world history, which spans more than 3000 years in Greece and Italy, over 1000 years in Iceland, and less than 200 years in Hawaii. Some 500 to 600 active volcanoes are recognized on Earth, and the number grows as eruptions occur on volcanoes that have been dormant since prehistoric times.

Volcanoes have been classified in several ways; by their shapes and sizes, by eruptive habits (some are much more dangerous than others), by rock type, and by location. A recent approach has been to classify them by their plate-tectonic settings. This has the advantage of reducing the classification to only three types, each of which is closely related to geodynamic processes on a global scale. The three types are subduction-zone volcanoes, which occur along convergent plate margins; rift volcanoes, along separating plate boundaries; and hot-spot volcanoes within the plates. The following three articles describe the nature of these tectonic regions and provide important insights into the differing root systems of the three volcanic clans.

Approximately 80 percent of the world's active volcanoes occur along subduction zones where one of the Earth's tectonic plates is thrust beneath another. The island arcs of the western Pacific and the high volcanic ranges of western North and South America nearly encircle the Pacific Ocean with a volcanic "Ring of Fire." Another zone extends from the Mediterranean Sea through Asia to Indonesia.

Japan is a good example of a subduction zone, with about 50 active volcanoes along parts of the four island arcs. The deep-sea trenches that form the actual plate boundaries are located about 200 kilometers on the Pacific Ocean side of the volcanoes. The volcanic belt occurs within the overlying plate along a zone where the underlying plate is 100 to 200 kilometers beneath the surface.

It is not clear whether magma is generated by the subduction process at this depth, or whether deep fractures formed in this region of the overlying plate allow partial melt from the upper mantle to ascend to the surface. Those who believe that subduction helps generate magma argue that shearing of converging plates provides frictional heating, and also that water-laden surface rocks subducted in the descending plate mix with hot rock at depth. Experiments have shown that water-rich rocks under high pressure melt at lower temperatures than their dry equivalents. Opponents argue that the descending plate is colder than normal rocks at this depth because of its thermal inertia, so melting would be inhibited. They also point out that from compositional evidence, the magma in subduction-zone volcanoes appears to have formed at depths of less than 100 to 200 kilometers. Whatever the

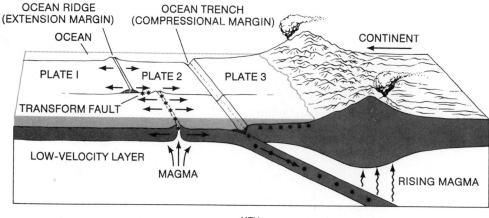

KEY

* SHALLOW EARTHQUAKES
(TENSION ON RIDGES; LATERAL SLIP ON TRANSFORM FAULTS)

• DEEP EARTHQUAKES
(MAINLY SHOWING THRUSTING AND DOWN DIP COMPRESSION)

Schematic cross-section of plate margins showing the relationships of earthquakes and volcanoes to ocean ridges (rift zones) and subduction zones. (After L. Sykes et al., in F. Press and R. Siever, *Earth*, first ed., p. 644. W. H. Freeman and Company, Copyright © 1974.)

cause, there is a very strong correlation between the location of active volcanoes and depth to the earthquake zone that marks the descending plate. Most major volcanoes along island arcs are about 50 to 70 kilometers apart. Their rock types are quite varied, from basalts (about 50 percent SiO_2) through andesites and dacites (55 percent to 65 percent SiO_2) to rhyolites (70 percent SiO_2). Most of their eruptive products are explosive fragments, with smaller volumes of lava flows. Strato-volcanoes composed of alternating layers of ash and lava are the predominant form of subduction volcanoes. Mount Fuji, the graceful, slightly concave-upward cone rising to 3776 meters above Tokyo, is a near-perfect example.

Rift volcanoes account for only about 15 percent of the world's known active volcanoes, and most of them are in Iceland and East Africa. However, the 75,000-kilometer-long world rift system is largely submarine. If the oceans were drained away, it would probably be found that rift volcanoes are the most common type. It has been estimated that there are about 20 eruptions of deep submarine rift volcanoes every year. The fact that not one of these eruptions has ever been witnessed, even by oceanographic instruments, creates an exploration challenge of the highest rank.

Most oceanographers and volcanologists believe that a shallow magma chamber exists along the axis of the world rift system. It apparently is replenished as hot plastic rock from the upper mantle rises to fill the void caused by the separating plates. The reduction in pressure on the ascending rock lowers its melting temperature and some of the heat in the rising hot plastic rock causes melting of the less refractory constituents of the mantle. Some of this magma escapes to the ocean floor along the rift axis, but much of it heals the scar of the diverging plates in the form of dikes and other shallow intrusions.

Nearly all deep submarine volcanic products are basalt lava flows, generally pillow basalts, which are sack-like bodies that expand from the inside until they split open to form new pillows. Basaltic volcanoes extruded in shallow water like Surtsey in Iceland, are explosive because of the rapid boiling of contact seawater, but once they rise above sea level they produce relatively quiet lava flows. Flows from long fissures parallel to the rift system are characteristic of rift volcanoes, but repeated flows from localized vents build gently sloping shield volcanoes. Rift volcanoes in continental settings such as East Africa produce a greater variety of rock types than their oceanic counterparts.

Active hot-spot volcanoes are the least common type, with Hawaii the principal example. They are located within the plates, and as a plate moves slowly over a hot spot, a chain of volcanoes that is progressively older in the direction of plate motion is formed. Their roots are enigmatic, but recent geochemical data indicate that deep mantle plumes may be their cause. Oceanic hot-spot volcanoes are basaltic; continental ones such as those in Yellowstone Park vary from basalt to rhyolite.

Support for the somewhat controversial concept of hot-spot volcanism is still growing, though some authorities prefer many fewer hot spots than the number shown on Burke and Wilson's diagram on page 37. The hot-spot theory may even explain why Iceland is the only major part of the Mid-Atlantic ridge that is above sea level. If Iceland is a hot spot that has become straddled by a rift system, then it has volcanoes of dual origin, which could account for the extra volcanism that has built Iceland above the sea.

Beneath the sea there is still much to be learned. There are many submarine volcanoes on the ocean floor that do not seem to be either of rifting or of hot-spot origin. We have no data on the volcanic activity of these seamounts; most of them are probably extinct. It is always wise to include an "Others" listing in any classification scheme.

Classifying volcanoes according to plate tectonics is the latest but probably not the last word. However, it does have the advantage of relating volcanoes to fundamental processes that affect the entire Earth. The historic activity of individual volcanoes encompasses such a short span of geologic time that their natures are difficult to ascertain; grouping them into clans increases the data base from which to generalize.

1

The Subduction of the Lithosphere

by M. Nafi Toksöz
November 1975

The rocky shell of the earth grows outward from mid-ocean ridges. Ultimately it plunges into the mantle below, giving rise to oceanic trenches, earthquakes, volcanoes, island arcs and mountain ranges

The lithosphere, or outer shell, of the earth is made up of about a dozen rigid plates that move with respect to one another. New lithosphere is created at mid-ocean ridges by the upwelling and cooling of magma from the earth's interior. Since new lithosphere is continuously being created and the earth is not expanding to any appreciable extent, the question arises: What happens to the "old" lithosphere?

The answer came in the late 1960's as the last major link in the theory of sea-floor spreading and plate tectonics that has revolutionized our understanding of tectonic processes, or structural deformations, in the earth and has provided a unifying theme for many diverse observations of the earth sciences. The old lithosphere is subducted, or pushed down, into the earth's mantle. As the formerly rigid plate descends it slowly heats up, and over a period of millions of years it is absorbed into the general circulation of the earth's mantle.

The subduction of the lithosphere is perhaps the most significant phenomenon in global tectonics. Subduction not only explains what happens to old lithosphere but also accounts for many of the geologic processes that shape the earth's surface. Most of the world's volcanoes and earthquakes, including nearly all the earthquakes with deep and intermediate foci, are associated with descending lithospheric plates. The prominent island arcs—chains of islands such as the Aleutians, the Kuriles, the Marianas and the islands of Japan—are surface expressions of the subduction process. The deepest trenches of the world's oceans, including the Java and Tonga trenches and all others associated with island arcs, mark the seaward boundary of subduction zones. Major mountain belts, such as the Andes and the Himalayas, have resulted from

the convergence and subduction of lithospheric plates.

In order to appreciate the gigantic scale on which subduction takes place, consider that both the Atlantic and the Pacific oceans were created over the past 200 million years as a consequence of sea-floor spreading. Thus the lithosphere that underlies the world's major oceans is less than 200 million years old. As the oceans opened, an equivalent area of lithosphere was simultaneously subducted. A simple calculation shows that the process involved the consumption of at least 20 billion cubic kilometers of crustal and lithospheric material. At the present rate of subduction an area equal to the entire surface of the earth would be consumed by the mantle in about 160 million years.

To understand the subduction process it is necessary to look at the thermal regime of the earth. The temperatures within the earth at first increase rapidly with depth, reaching about 1,200 degrees Celsius at a depth of 100 kilometers. Then they increase more gradually, approaching 2,000 degrees C. at about 500 kilometers. The minerals in peridotite, the major constituent of the upper mantle, start to melt at about 1,200 C., or typically at a depth of 100 kilometers. Under the oceans the upper mantle is fairly soft and may contain some molten

material at depths as shallow as 80 kilometers. The soft region of the mantle, over which the rigid lithospheric plate normally moves, is the asthenosphere. It appears that in certain areas convection currents in the asthenosphere may drive the plates, and that in other regions the plate motions may drive the convection currents.

The mid-ocean ridges mark the region where upwelling material forms new lithosphere. The ridges are elevated more than three kilometers above the average level of the ocean floor because the newly extruded rock is hot and hence more buoyant than the colder rock in the older lithosphere. As the lithosphere spreads away from the ridge it gradually cools and thickens. The spreading rate is generally between one centimeter and 10 centimeters per year. The higher velocities are associated with the Pacific plate and the lower velocities with the plates bordering the Mid-Atlantic Ridge. At a velocity of eight centimeters per year the lithosphere will reach a thickness of about 80 kilometers at a distance of 1,000 kilometers from the ridge. Under most of the Pacific abyssal plains a thickness of this value has been confirmed by measurements of the velocities of seismic waves.

Where two plates move toward each

HIMALAYAS OF NEPAL, shown in the false-color picture (opposite page) made from the Earth Resources Technology Satellite (ERTS), are a zone in which continental lithosphere is being subducted. In most subduction zones oceanic lithosphere plunges under continental lithosphere. Here the lithosphere of the Indian subcontinent (*bottom*) is being subducted under the snow-covered Himalayas (*top*), raising the mountain range in the process. The area covered by picture is 125 kilometers (78 miles) across. Mount Everest is one of the peaks on the ridge at the very edge of the picture in the upper right-hand corner. The main boundary fault between the two lithospheric plates runs from left to right in the valley that is marked by two clusters of cloud that are visible at lower center and lower right.

other and converge, the oceanic plate usually bends and is pushed under the thicker and more stable continental plate. The line of initial subduction is marked by an oceanic trench. At first the dip, or angle of descent, is low and then it gradually becomes steeper. Profiles across trenches of the reflections of seismic waves clearly show the downward curve of the top of descending oceanic plates.

Several factors contribute to the heating of the lithosphere as it descends into the mantle. First, heat simply flows into the cooler lithosphere from the surrounding warmer mantle. Since the conductivity of the rock increases with temperature, the conductive heating becomes more efficient with increasing depth. Second, as the lithospheric slab descends it is subjected to increasing pres-

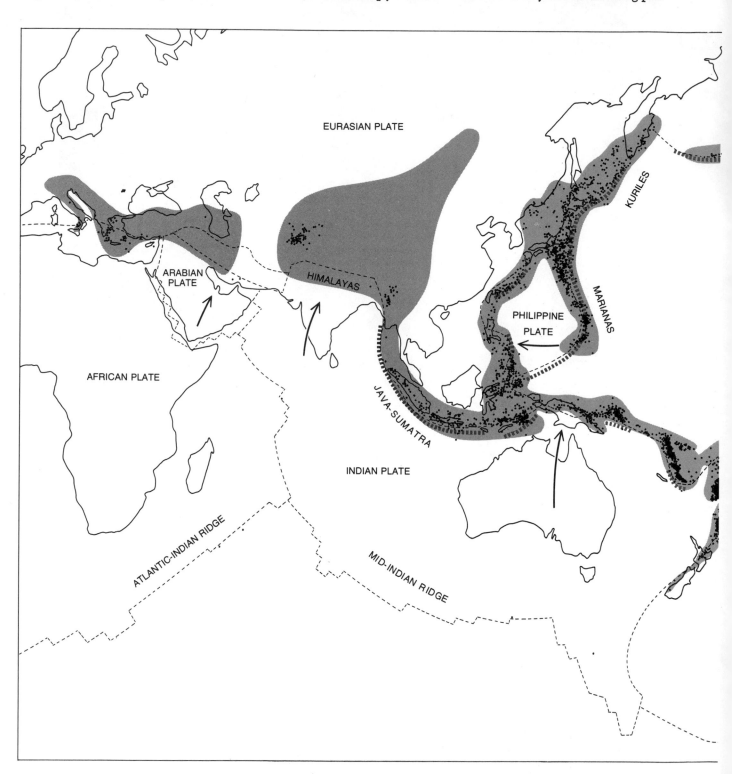

TECTONIC MAP OF THE EARTH depicts the principal lithospheric plates and their general direction of motion (*arrows*). New material is continually added to the plates at mid-ocean ridges by the upwelling and cooling of magma from the earth's mantle. It moves outward and is eventually returned to the mantle by sub-duction. There it is slowly consumed. The subduction process creates deep oceanic trenches (*broken lines in color*) and island arcs such as those bordering the western and northern Pacific. On the islands of the arcs are many active volcanoes. Young mountain belts in Europe and Asia identify zones where continental litho-

sure, which introduces heat of compression. Third, the slab is heated by the radioactive decay of uranium, thorium and potassium, which are present throughout the earth's crust and add heat at a constant rate to the descending material. Fourth, heat is provided by the energy released when the minerals in the lithosphere change to denser phases, or more compact crystal structures, as they are subjected to higher pressures during descent. Finally, heat is generated by friction, shear stresses and the dissipation of viscous motions at the boundaries between the moving lithospheric plate and the surrounding mantle. Among all these sources the first and fourth contribute the most toward the heating of the descending lithosphere.

The temperatures inside a descending

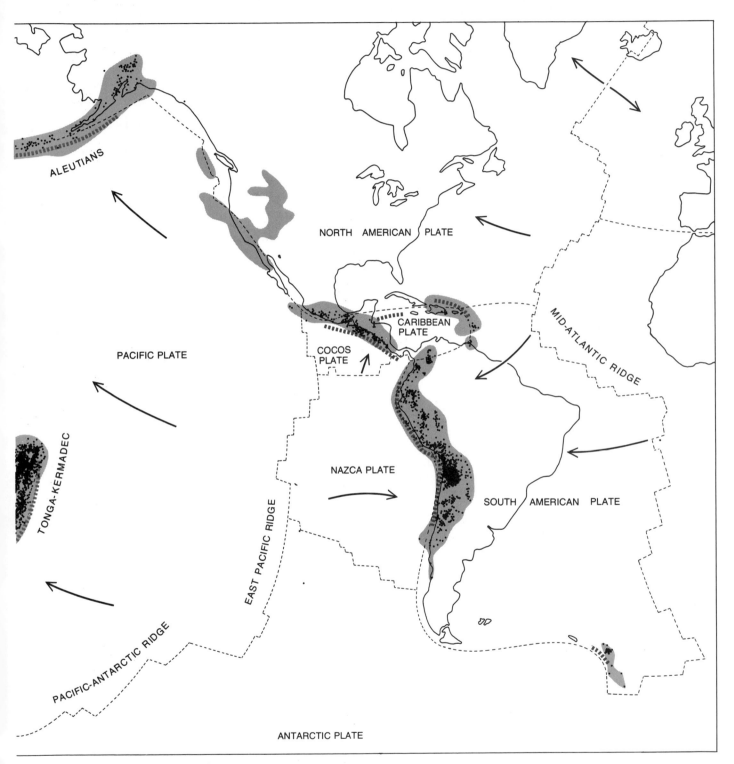

spheric plates converge; around the Pacific young mountain ranges result from the subduction of oceanic plates. The areas in color identify the general location of the great majority of earthquakes that occurred at all depths between 1961 and 1967; they are based on maps made by H. J. Dorman and M. Barazangi of the Lamont-Doherty Geological Observatory. Most earthquakes have a magnitude below 6.5 and occur at a shallow depth (between five and 15 kilometers). The locations of deep earthquakes, those occurring below 100 kilometers, are given by the black dots. All the deep earthquakes take place in cold descending slabs of the oceanic type.

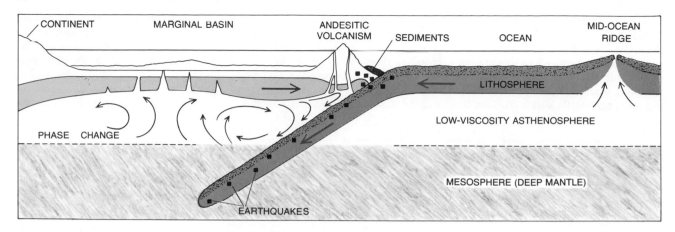

FORMATION AND SUBDUCTION OF LITHOSPHERE are shown in this cross section of the crust and mantle. New lithosphere is created at a mid-ocean ridge. A trench forms where the lithospheric slab descends into the mantle. Earthquakes (*small squares*) occur predominantly in the upper portion of the descending slab. Arrows in soft asthenosphere indicate direction of possible convective motions. Secondary convection currents in asthenosphere may form small spreading centers under marginal basins.

lithospheric plate have been calculated theoretically over the past five years by geophysicists in Britain, Japan and the U.S. Although different approaches were taken in the calculations, the results are in good agreement. For example, our group at the Massachusetts Institute of Technology has computed the progressive heating of plates penetrating into the mantle at various velocities over periods ranging from several hundred thousand years to more than 10 million years. A typical calculation based on our model of the phenomenon shows what happens to a plate descending at the rate of eight centimeters per year (the velocity characteristic of the Pacific subduction zones) at three points in time: 3.6, 7.1 and 12.4 million years after the beginning of subduction [*see illustration on opposite page*].

NAME	PLATES INVOLVED	TYPE	LENGTH OF ZONE (KILOMETERS)	SUBDUCTION RATE (CENTIMETERS PER YEAR)	MAXIMUM EARTHQUAKE DEPTH (KILOMETERS)	TYPE OF SUBDUCTING LITHOSPHERE
KURILES-KAMCHATKA-HONSHU	PACIFIC UNDER EURASIAN	A	2,800	7.5	610	OCEANIC
TONGA-KERMADEC-NEW ZEALAND	PACIFIC UNDER INDIAN	A	3,000	8.2	660	OCEANIC
MIDDLE AMERICAN	COCOS UNDER NORTH AMERICAN	B	1,900	9.5	270	OCEANIC
MEXICAN	PACIFIC UNDER NORTH AMERICAN	B	2,200	6.2	300	OCEANIC
ALEUTIANS	PACIFIC UNDER NORTH AMERICAN	B	3,800	3.5	260	OCEANIC
SUNDRA-JAVA-SUMATRA-BURMA	INDIAN UNDER EURASIAN	B	5,700	6.7	730	OCEANIC
SOUTH SANDWICH	SOUTH AMERICAN SUBDUCTS UNDER SCOTIA	C	650	1.9	200	OCEANIC
CARIBBEAN	SOUTH AMERICAN UNDER CARIBBEAN	C	1,350	0.5	200	OCEANIC
AEGEAN	AFRICAN UNDER EURASIAN	C	1,550	2.7	300	OCEANIC
SOLOMON–NEW HEBRIDES	INDIAN UNDER PACIFIC	D	2,750	8.7	640	OCEANIC
IZU-BONIN-MARIANAS	PACIFIC UNDER PHILIPPINE	D	4,450	1.2	680	OCEANIC
IRAN	ARABIAN UNDER EURASIAN	E	2,250	4.7	250	CONTINENTAL
HIMALAYAN	INDIAN UNDER EURASIAN	E	2,400	5.5	300	CONTINENTAL
RYUKYU-PHILIPPINES	PHILIPPINE UNDER EURASIAN	E	4,750	6.7	280	OCEANIC
PERU-CHILE	NAZCA UNDER SOUTH AMERICAN	E	6,700	9.3	700	OCEANIC

MAJOR SUBDUCTION ZONES and some of their principal characteristics are listed. One of the smallest plates, the Nazca plate, is associated with the longest single subduction zone, embracing almost the entire west coast of South America. It also has the second-highest subduction rate: 9.3 centimeters per year perpendicular to the arc of the earth's surface. In general the more rapidly a plate descends, the greater is the maximum depth of earthquakes associated with it. (A major exception is the subduction zone under the Philippines.) The five principal types of subduction zone (A–E) are depicted schematically in the illustration on page 12.

In this model the interior of the descending plate remains distinctly cooler than the surrounding mantle until the plate reaches a depth of about 600 kilometers. As the plate penetrates deeper its interior begins to heat up more rapidly because of the more efficient transfer of heat by radiation. When the plate goes beyond a depth of about 700 kilometers, it can no longer be thermally distinguished as a structural unit. It has become a part of the mantle. Significantly, 700 kilometers is a depth below which no earthquake has ever been recorded. Apparently deep earthquakes cannot occur except in descending plates; therefore the occurrence of such earthquakes implies the presence of sunken plate material.

The descending lithosphere does not always, however, penetrate to 700 kilometers before it is assimilated. A slow-moving plate will attain thermal equilibrium before reaching that depth. For example, at a velocity of one centimeter per year the subducting plate will be assimilated at a depth of about 400 kilometers. If subduction ceases altogether, the subducted segment of the lithosphere will lose its identity and become part of the surrounding mantle in roughly 60 million years. At half that age a stationary plate will already have become too warm to generate earthquakes. These calculations make it clear why we can identify only those subducted plates that are associated with the latest episode of sea-floor spreading. Although there are surface geological expressions of older subduction zones, the plates subducted under these regions cannot be identified in the earth's mantle. The old slabs are lost not only because of the assimilation process but also because of the motion of the surface with respect to the mantle.

So far I have been describing ideal subduction zones without major complications. Such zones are found, for example, under the Japanese island of Honshu, under the Kuriles (extending to the north of Japan) and under the Tonga-Kermadec area (to the north of New Zealand). In many other areas the lithosphere descends in a more complicated manner.

In new subduction areas the descending slab may have penetrated a good deal less than 700 kilometers, as is the case under the Aleutians, the west coast of Central America and Sumatra. In other areas where the subduction rate is low the slab may be assimilated well before it reaches that depth; the subduction

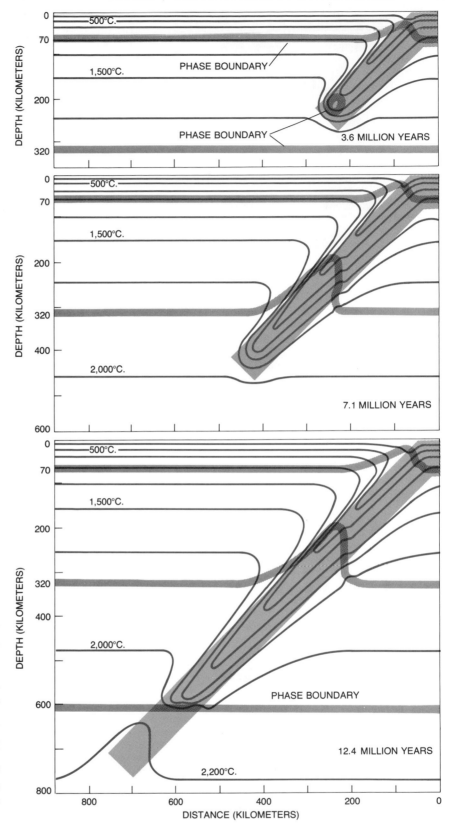

EVOLUTION OF DESCENDING SLAB is described by computer models developed in the author's laboratory at the Massachusetts Institute of Technology. These diagrams depict the fate of a slab subducting at an angle of 45 degrees and at a rate of eight centimeters per year. Phase changes, induced by increasing pressure, normally occur at depths of 70, 320 and 600 kilometers. In the descending slab the first two phase changes occur at shallower depths because of the slab's lower temperature. The phase conversions to denser mineral forms help to heat the slab and to speed its assimilation. When the slab reaches the temperature of the surrounding mantle at a depth of 700 kilometers, it loses its original identity.

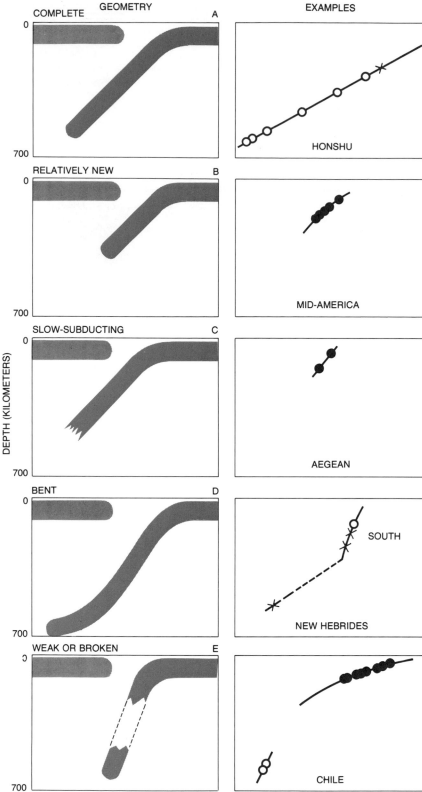

FIVE MAJOR TYPES of subducting oceanic slabs can be identified. Examples of each type are shown to the right of the schematic diagrams. In the examples the solid lines represent the location of all earthquakes projected onto a cross section. The symbols on the lines identify particularly large earthquakes from which the direction of stress was determined. Open circles indicate compression along the length of the slab; filled circles indicate tension along the length of the slab, and crosses show stresses that do not lie in the plane of the cross section. Many subduction zones exhibit a "seismic gap" between 300 and 500 kilometers where no earthquakes occur. It is not known whether this is because the slab is broken (*Type E*) or because stresses are absent at that depth. Examples given are based on a survey conducted by Bryan L. Isacks of Cornell University and Peter Molnar of M.I.T.

of the Mediterranean plate under the Aegean Sea is an example. In still other areas the subduction starts at a shallow angle, gets steeper at intermediate depths and bends again nearly to the horizontal at about 500 kilometers. Such a sigmoid configuration is observed dramatically under the New Hebrides in the South Pacific. The double bend may be attributable to low resistance in the upper asthenosphere and much greater resistance at a depth of 600 kilometers, resulting from an increase either in the density or in the strength of the mantle, possibly both. Another anomalous situation is found under Peru and Chile, where there is a marked absence of earthquakes at intermediate depths, indicating a stress-free zone or possibly a broken slab.

Most frequently the oceanic lithosphere is subducted under an island arc, as is generally the case in the western Pacific. Here, however, there are many other combinations and complications. For example, a small oceanic plate, such as the Philippine plate, may get trapped between two trenches. Or an oceanic plate may be subducted under a continent, as in the case of the Nazca plate, which plunges under the Andes. The Andes can be regarded as being equivalent to an overgrown island arc. Elsewhere transform faults such as the San Andreas fault may interrupt subduction boundaries. In other cases multiple subduction zones may develop within relatively small areas. Finally, subducting plates may bring two continents together, with major tectonic consequences. Continental collisions place major restrictions on plate motions because the buoyancy of the continental crust, which is less dense than the mantle, resists subduction. Collisions of this type create major mountain belts, such as the Alps and the Himalayas.

Continental subduction is qualitatively different from oceanic subduction because it is a transient process rather than a steady-state one. When continental crust moves into a subduction zone, its buoyancy prevents it from being carried down farther than perhaps 40 kilometers below its normal depth. As plate convergence continues the crust becomes detached from the plate and is itself underthrust by more continental crust. That creates a double layer of low-density crust, which rises buoyantly to support the high topography of a major mountain range. It is possible that the long oceanic slab below the surface ultimately becomes detached and sinks; in any case it is no longer a source of earthquakes. After this stage further deformation and compression may take place behind the line of collision, producing a high plateau with surface volcanoes, like the plateau of Tibet. Eventually the plate convergence itself will stop as resisting forces build up. It now seems that continental collisions are probably a major factor in the periodic reorientations of the relative motions of the plates.

It is clear that an understanding of the geological, geochemical and geophysical consequences of lithospheric subduction helps to explain many major features of the earth's surface. At the same time the observable features enable us to test the validity of theoretical subduction models. A wide variety of features can be investigated. For the sake of brevity I shall mention only the geological characteristics of the trench sediments and the subducting crust, the andesitic magmas associated with island-arc volcanoes, and heat-flow and gravity anomalies. The measurable quantities related to these features are primarily sensitive to the properties of subduction down to a depth of about 100 kilometers. The most definitive observations on the deeper parts of subducting plates are seismic observations. The velocity and attenuation of seismic waves, and most significantly the indication the waves give of the locations of deep- and intermediate-focus earthquakes, outline the extent of the relatively cool and rigid zone of the descending lithosphere.

With the passage of time the deep oceanic trenches created by descending plates accumulate large deposits of sediment, primarily from the adjacent continent. As the sediments get caught between the subducting oceanic crust and either the island arc or the continental crust they are subjected to strong deformation, shearing, heating and metamorphism. Profiles of seismic reflections have identified these deformed units. Some of the sediments may even be dragged to great depths, where they may eventually melt and contribute to volcanism. In this case they would return rapidly to the surface, and the total mass of low-density crustal rocks would be preserved.

A prominent feature of subduction zones is volcanism that gives rise to andesite, a fine-grained gray rock. Where the magma for these volcanoes originates is not definitely known. Most geochemical and petrological evidence favors a depth of about 100 kilometers for the magma source. The magma may come

from the partial melting of the subducted oceanic crust, as A. E. Ringwood of the Australian National University suggested in 1969. The shearing that takes place at the top of the descending plate may provide the heat required for partial melting. Convective motions in the wedge of asthenosphere above the descending plate may also contribute to magma sources by raising asthenospheric material to a depth where it could melt slightly under lower pressure.

The flow of heat through the earth's surface tells us something about the thermal characteristics of shallow layers. (It is influenced only indirectly by deeper

phenomena.) Trenches have low heat flow (less than one microcalorie per square centimeter per second); island arcs generally have a high and variable heat flow because of their volcanism. High heat flow is also associated with the marginal basins behind the island arcs, for example the Sea of Japan, the Sea of Okhotsk, the Lau Basin west of Tonga and the Parece Vela Basin behind the Marianas arc.

These basins are underlain by relatively hot material brought up either by convection currents behind the island arc or by upwelling from deeper regions. The convection is induced in the wedge

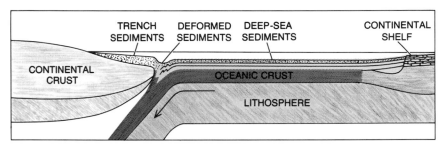

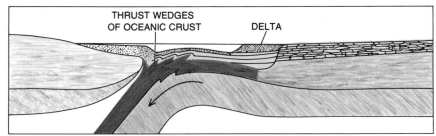

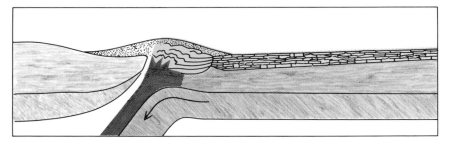

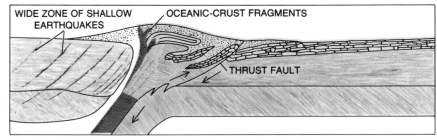

COLLISION OF CONTINENTS occurs when an oceanic slab that is subducting at the edge of one continent (*left*) is itself part of a lithospheric plate bearing a second continent (*right*). Such a collision took place when the Indian lithospheric plate, traveling generally northward for 200 million years, subducted under Eurasian plate. This kind of subduction eventually ends, but not before crust of subducting plate has been detached and deformed and has pushed up a mountain range (in the case of the Indian plate the Himalayas).

of asthenosphere above the descending lithospheric plate by the downward motion of the plate. Since it takes time for such currents to be set in motion, high heat flow would not be expected in basins behind the youngest subduction zones. Indeed, the observed heat-flow values in the Bering Sea behind the Aleutians are normal.

Gravity anomalies associated with subduction zones are large and broad. A descending lithospheric plate is cooler and denser than the surrounding mantle; therefore it gives rise to a positive gravity anomaly. The hot region under a marginal basin would show a density lower than normal and hence would create a negative gravity anomaly. Changes in the character of the crust from the ocean to an island arc or a continent add more anomalies. A combination of all these anomalies is needed to account for the gravity observations that have been made across subduction zones [*see illustration on opposite page*]. The gravity evidence provides strong support for the

subduction models, but it is not conclusive because of the uncertainties as to the depth of the masses that give rise to the anomalies.

The most compelling evidence for the subduction of the lithosphere comes from seismology. Most of the world's earthquakes and nearly all the deep- and intermediate-focus earthquakes are associated with subduction zones. The hypocenters of the earthquakes and their source mechanisms can be explained by the stresses in the subducting plate. The models that explain the seismic-wave observations outline the location of the subducted cool lithospheric plates. In some areas (Japan, the Aleutians, the Tonga Trench, South America) the data are abundant and convincing.

The general picture that emerges is as follows. At shallow depths, where the edges of the two rigid lithospheric plates are pressing against each other, there is intense earthquake activity. Many of the world's greatest earthquakes (for example the Chile earthquake of 1960, the

Alaska earthquake of 1964 and the Kamchatka earthquake of 1952), as well as many smaller ones, occur along the shear plane between the subducting oceanic lithosphere and the continental or island-arc lithosphere. Some normal-faulting (tensional) earthquakes on the ocean side of a trench are caused by arching of the lithosphere. Other earthquakes result from the tearing of the lithosphere and other adjustments in this zone of intense deformation.

The deep- and intermediate-focus earthquakes generally occur along the Benioff zone, a plane that dips toward a continent. At first this plane was thought to be the shear zone between the upper surface of the descending lithospheric plate and the adjoining mantle. Detailed studies conducted by Bryan L. Isacks of Cornell University and Peter Molnar of M.I.T. and others over the past 10 years have shown, however, that the forces needed to account for the observed earthquakes could not be provided by the shearing process. These studies, combined with more precise determinations of the location of earthquake foci under several island arcs, indicate that the deep- and intermediate-focus earthquakes occur in the coolest region of the interior of the descending plate. The stresses generated by the gravitational forces acting on the dense interior of the slab and the resistance of the surrounding mantle to the slab's penetration are also highest in the coolest region. Moreover, the cool and rigid interior of the slab acts as a channel to transmit stresses. The computed directions of the stresses are consistent with the directions that have been deduced from earthquakes.

These concepts can be tested in areas where detailed studies of earthquakes have been made. Two such regions are the Aleutians and Japan. At Amchitka Island in the central Aleutians the nuclear explosions named Longshot, Milrow and Cannikin provided energy sources with precisely known locations and times. From the travel times of the seismic waves going through the subducting lithosphere the location of the coolest region was determined precisely. The dense network of seismic stations installed in the area also provided precise locations of earthquakes. The shallow earthquakes are concentrated along the thrust plane and the deeper ones along the coolest region [*see illustration at left*].

The islands of Japan constitute probably the most intensively studied seismic belt in the world. The velocities of seis-

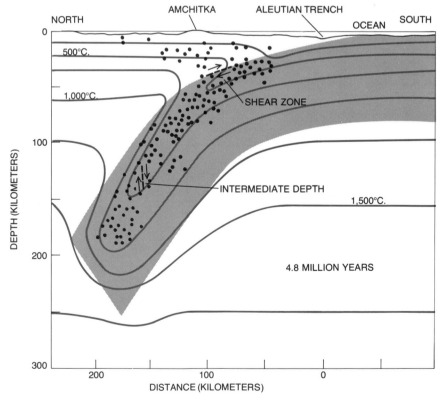

ALEUTIAN EARTHQUAKES mark the general location of the subducting Pacific plate in that region. The precise location of the cold descending slab in relation to the earthquakes was determined with the help of seismic waves from nuclear tests on Amchitka Island, which showed that the waves travel more rapidly through the cold slab than through the surrounding mantle. A computed simulation of seismic records revealed that intermediate-depth earthquakes (*dots*) occur in the cold center of the slab, as shown here, and not, as had been thought, at the shear zone on its upper face. At shallower depths the earthquakes occur in shear zone and in overriding plate. Arrows show the slip planes and the sense of motion.

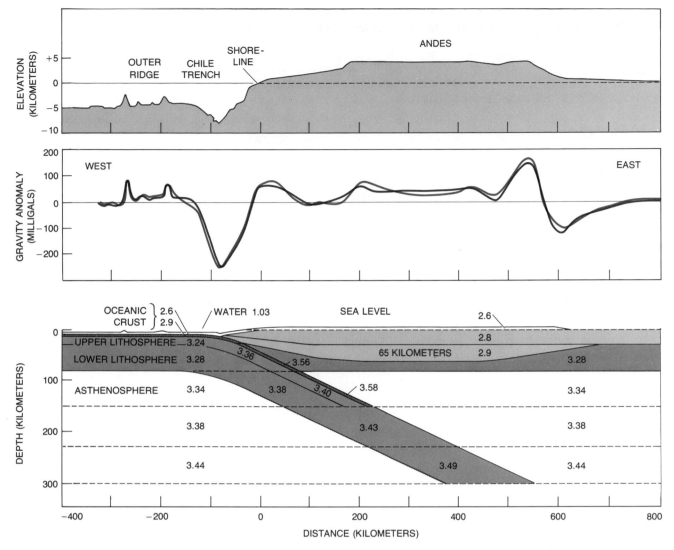

EFFECT OF A SUBDUCTING PLATE ON GRAVITY is clearly represented in the gravity anomaly that has been measured over the west coast of Chile and the Andes. The diagram at the top is a topographical cross section of the region. The observed gravity anomaly, given in milligals, is shown by the black curve in the middle diagram. The colored curve is the anomaly calculated on the basis of the lithospheric model shown at the bottom. (One gal, named for Galileo, is one 980th the normal gravity at the earth's surface; thus an anomaly of −260 milligals over the trench corresponds to a gravity deficit of about .026 percent.) The model includes the trench, which gives rise to the gravity low, and cold dense slab, which has opposite effect. Densities given in model are in grams per cubic centimeter. Model was worked out by J. A. Grow and Carl O. Bowin of the Woods Hole Oceanographic Institution.

mic waves, the characteristics of the waves' attenuation, the precise locations of earthquake hypocenters and the focal mechanisms all fit the subduction model in this region. The descending plate shows high velocities and low attenuation, which is a measure of the nonelastic damping of high-frequency seismic waves. There are numerous shallow earthquakes along and near the boundary where the plates meet near the surface. Deep- and intermediate-focus earthquakes are in the coolest region of the slab where the stresses are highest [see illustration on next page]. In other subduction zones the locations of earthquakes are not as precisely known. Nevertheless, wherever adequate data exist, for example for the areas of the Tonga Trench and of Peru and Chile, the deep- and intermediate-focus earthquakes are found to occur in the interior of the subducting plate along the coolest region.

The absence of earthquakes below a depth of 700 kilometers can now be explained. The descending lithosphere heats up below that depth and can no longer behave as a rigid elastic medium susceptible to faulting or brittle fracture. Moreover, below that depth the stresses are small, and they are relieved by slow plastic deformation rather than by the sudden failure associated with an earthquake.

The gravitational energy associated with large masses of subducting cool, dense material is large even in terms of the total energy associated with plate motions. The gravitational forces are largely balanced by the resistance of the mantle to the penetration of the descending lithosphere. The net force acting on the plates in the subduction zone is still enough to play a major role in global plate motions. Other forces that contribute are the horizontal flow of convection currents under the plates and the outward push of the material coming to the surface at the mid-ocean ridges.

Not all the problems of plate motions and subduction have been solved. It is puzzling, for example, that the Pacific plate can move laterally for 6,000 kilometers before it subducts. It is not known why some subduction zones are where they are. It is not clear why plate motions change at certain times. These are minor problems, however, compared with the understanding of continental drift, earthquakes, volcanism and mountain building that has been gained. The theory of plate tectonics is a concept that unifies the main features of the earth's surface and their history better than any other concept in the geological sciences.

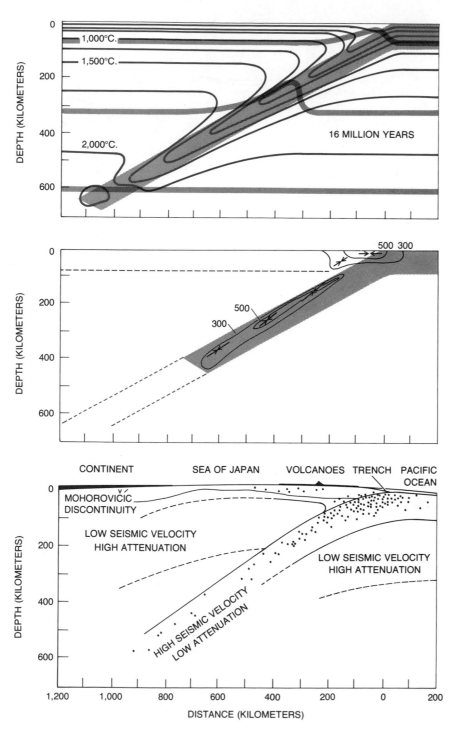

EARTHQUAKES IN JAPAN AREA are caused by several westward-dipping slabs of Pacific lithosphere. The author has calculated a temperature model for a typical Japan slab (*top*). This in turn has been used to calculate the stresses generated within the upper portion of the slab (*middle*). The stresses result from the interaction between the slab's tendency to sink because of its high density and opposing forces: friction near surface and viscous drag in asthenosphere. Nonhydrostatic stresses are computed in bars (one bar is 14.7 pounds per square inch). Arrows show direction of compression. Calculated stresses account well both for distribution of earthquakes (*bottom*) and their mode of initiation.

2

The Crest
of the East Pacific Rise

by Ken C. Macdonald
and Bruce P. Luyendyk
May 1981

*At a site on a mid-ocean ridge, where hot springs
on the sea floor nourish a bizarre biological community,
undersea exploration has revealed much about how
new segments of the earth's crust emerge*

The East Pacific Rise is a part of the world's longest mountain chain: the 75,000-kilometer oceanic rift system, which winds around the globe like the seam of a baseball. Most of the rift system, including the entire East Pacific Rise, lies underwater; it is a network of mid-ocean ridges. In the plate-tectonic theory that has become the organizing principle of geology, the rift system has a vital role. Each rift is a narrow, fractured zone where plates of the oceanic crust are continually being pushed or pulled apart. At such a sea-floor spreading center molten rock wells up from the mantle of the earth, filling cracks and generating new sections of oceanic crust, which move outward as if they were on a broad conveyor belt.

The concept of sea-floor spreading accounts for a host of geologic observations. Still, several important questions about the rift system remain unanswered. Underwater photographs and rock samples show signs of recent volcanic activity on the mid-ocean ridges, suggesting there is a magma chamber, or reservoir of molten rock, under the axis of the ridge. Does such a magma chamber exist as a permanent feature of the ridge? If it does, how deep is it? How far from the axis does it extend? What are the physical and chemical properties of the magma? How much heat is lost from the earth's interior through the volcanic creation of new crust? The answers to these questions and others like them could bring a much improved understanding of the structure and composition of the earth's crust, since at least 70 percent of the crust has formed at the mid-ocean rifts.

We and a group of our collaborators have recently had an opportunity to explore at close range a small region along the crest of the East Pacific Rise. From the expedition has come important new evidence of a magma chamber along the axis of the crest and of a narrow, sharply defined zone of crustal formation above the chamber. The evidence is derived from measurements of seismic, electrical, gravimetric and magnetic properties

of the crust and detailed geologic maps made along traverses of the axial volcanic zone. The most dramatic evidence comes from direct visual observations made with a small submersible vehicle at a depth of more than two and a half kilometers. At several places along the crest we found clusters of hydrothermal vents spewing out mineral-blackened water that had been heated to extraordinarily high temperatures by contact with rocks near a magma chamber. It turns out that the hydrothermal vents have a major influence not only on the geophysics of the rift system but also on the chemical balance of the oceans. Furthermore, the vents support an unusual biological community, the only one known that is entirely independent of photosynthetic sources of energy.

Plate Tectonics

The theory of plate tectonics describes the motions of the lithosphere, the comparatively rigid outer layer of the earth that includes all the crust and part of the underlying mantle. The lithosphere is divided into a few dozen plates of various sizes and shapes; in general the plates are in motion with respect to one another. A mid-ocean ridge is a boundary between plates where new lithospheric material is injected from below. As the plates diverge from a mid-ocean ridge they slide on a more yielding layer at the base of the lithosphere.

Since the size of the earth is essentially constant, new lithosphere can be created at the mid-ocean ridges only if an equal amount of lithospheric material is consumed elsewhere. The site of this destruction is another kind of plate boundary: a subduction zone. There one plate dives under the edge of another and is reincorporated into the mantle. Both kinds of plate boundary are associated with fault systems, earthquakes and volcanism, but the kinds of geologic activity observed at the two boundaries are quite different.

The idea of sea-floor spreading actually preceded the theory of plate tecton-

ics. The sea-floor-spreading hypothesis was formulated chiefly by Harry H. Hess of Princeton University in the early 1960's. In its original version it described the creation and destruction of ocean floor, but it did not specify rigid lithospheric plates. The hypothesis was substantiated soon afterward by the discovery that periodic reversals of the earth's magnetic field are recorded in the oceanic crust. An explanation of this process devised by F. J. Vine and D. H. Matthews of Princeton is now generally accepted. As magma rises under the mid-ocean ridge, ferromagnetic minerals in the magma become magnetized in the direction of the geomagnetic field. When the magma cools and solidifies, the direction and the polarity of the field are preserved in the magnetized volcanic rock. Reversals of the field give rise to a series of magnetic stripes running parallel to the axis of the rift. The oceanic crust thus serves as a magnetic tape recording of the history of the geomagnetic field. Because the boundaries between stripes are associated with reversals of the magnetic field that can be dated independently, the width of the stripes indicates the rate of sea-floor spreading. (Precisely how the earth's magnetic field reverses at intervals of from 10,000 to about a million years continues to be one of the great mysteries of geology.)

It follows from the theory of sea-floor spreading that many of the most interesting geologic features of the earth's surface are to be found on the ocean floor. The investigation of such features has been furthered in recent years by the development of deep-diving manned submersibles. In particular the U.S. research submersible *Alvin,* operated by the Woods Hole Oceanographic Institution, has proved to be a valuable tool for studies of the sea bed. A geologist in the *Alvin* can collect rock samples and document in detail the setting of each rock. For the first time a marine geologist can have maps of a site as precise as those of a geologist on land.

Early work with the *Alvin,* beginning in 1973, made it clear that submersibles

HOT, SULFIDE-BLACKENED WATER spews out of a "black-smoker chimney" at the crest of the East Pacific Rise near the entrance to the Gulf of California. The photograph was made from the manned research submersible *Alvin* at a depth of 2,650 meters; a part of the craft's sampling equipment is visible in the foreground. The East Pacific Rise is a boundary where two plates of the earth's lithosphere (a layer that includes the rocky crust) are being pulled apart. At the plate boundary cold seawater seeps into the comparatively thin crust and is heated by contact with magma, or molten rock, under the sea-floor spreading center. The water issues from the vents at temperatures as high as 350 degrees Celsius, bearing minerals that nourish a biological community. The site was explored in 1979 by a dive team that included the authors and 19 other investigators from the U.S., France and Mexico. The other members of the team were Fred N. Spiess, Charles S. Cox, James W. Hawkins, Rachel Haymon, J. Douglas Macdougall, John A. Orcutt, Loren Shure, Tanya M. Atwater, Stephen P. Miller, Robert D. Ballard, Jean Francheteau, T. Juteau, C. Rangin, R. Larson, William R. Normark, A. Carranza, V. M. Diaz Garcia, D. Cordoba and J. Guerrero. The photograph was made by Dudley Foster of the Woods Hole Oceanographic Institution.

are most efficiently deployed in the final stages of an undersea expedition. Time on the bottom is short (six hours or less) and expensive. Every means of mapping the sea floor, including the use of remotely operated cameras, high-resolution sonar scanners and other sensing devices, should be fully exploited so that the *Alvin* can be guided to key geologic sites. When the *Alvin* is employed in this sparing manner, it is a highly productive tool for gathering information. In the past seven years it has been used to investigate cycles of volcanic activity and patterns of geomagnetic reversal along the Mid-Atlantic Ridge and to study exposed cross sections of the crust in the Cayman Trough rift system near Jamaica. Hydrothermal vents and the associated exotic life forms were first observed in 1977 during dives made with the *Alvin* along the Galápagos spreading center off the coast of Ecuador.

We shall report here the results of the latest expedition with the *Alvin:* to the crest of the East Pacific Rise some 3,000 kilometers northwest of the Galápagos dive site. With the aid of the submersible and with instruments towed by surface vessels we measured such properties as the magnetization and the electrical conductivity of the crustal rocks, the velocity of seismic waves under the rise and the magnitude of gravitational anomalies over it. All these properties are sensitive indicators of the characteristics of the axial magma chamber thought to underlie the rift. It was during this expedition that we discovered the submarine hot springs that are the hottest yet found in the ocean.

Preliminary Explorations

The dive site we selected is near the northern end of the East Pacific Rise, just outside the entrance to the Gulf of California. The rise itself continues north up the middle of the gulf and links up with the San Andreas fault system in California. Outside the gulf it forms part of the boundary between the Pacific Plate and the Rivera Plate; the latter is a fragment of the much larger North American Plate.

The rift in the area we studied is currently spreading at a rate of some six centimeters per year, which is about as fast as human fingernails grow. The spreading rate here is three times the rate at the Mid-Atlantic Ridge, but it is only about a third of the highest spreading rate known: 18 centimeters per year, observed near Easter Island on another part of the East Pacific Rise. The site was picked because it was expected to be a typical intermediate-rate spreading center and because a considerable amount of detailed information about it was already available.

On earlier cruises we had obtained a fairly clear picture of the geologic setting of the spreading center and of its overall dimensions. Magnetic, photographic and sonar studies had been done with the aid of an unmanned vehicle, the Deep-Towed Instrument Package of the Scripps Institution of Oceanography. The studies indicated that at this point the spreading center may be only a kilometer or two wide. Bathymetric and geologic maps of the spreading center were assembled and targets for exploration were identified. It was established that the average depth of the dives would be more than 2,600 meters.

The first stage of the diving program was conducted in 1978 by a team of investigators from France, the U.S. and Mexico led by Jean Francheteau of the Center for Oceanographic and Marine Biological Studies in Brittany. Diving in the French submersible *Cyana*, the 12-member team focused on geologic investigations that called for the ability to maneuver near and visually reconnoiter rocky outcrops and other structures along the crest of the rise. Meanwhile planning proceeded for the next stage of the operation: the geophysical experiments, which were scheduled to begin in 1979 with the larger and stabler submersible *Alvin*.

The *Cyana* divers found that the spreading center actually consists of four geologic zones. Zone 1, directly on the axis of the spreading center, is a very young volcanic region approximately a kilometer wide. Almost all the new volcanic material produced at the spreading center appears to be extruded onto the sea floor within this remarkably narrow band. The basaltic lava flows found here, which are mostly in the form of the "pillows" characteristic of underwater eruptions, have essentially no sediments covering them. They exhibit a fresh, glassy luster and are comparatively unaltered by interactions with seawater.

Just outside the neovolcanic zone the newly formed crust begins to accelerate horizontally, approaching a maximum spreading velocity of three centimeters per year on each side of the spreading axis. In this region, designated Zone 2, the crust is stretched and cracked. Minor fissures tend to be lined up parallel to the general northeast trend of the rise and perpendicular to the direction of spreading. The fissured zone on each side of the central axis is between one-half kilometer and two kilometers wide.

Beyond Zone 2 the crust is probably still undergoing some acceleration, although in the next region, designated Zone 3, major "normal" faults begin to develop. The faults are almost vertical and look like a huge stairway. They are caused by abrupt vertical movements in rocks where the principal stress is tensional. Slippage along the faults gives rise to frequent earthquakes with a maximum magnitude of roughly 5.5 on the Richter scale. The scarps, or exposed faces of the faults, are generally oriented toward the spreading center and are as much as 70 meters high. At approximately 10 kilometers from the spreading center, in the region designated Zone 4, active faulting diminishes sharply, and so presumably does horizontal acceleration of the crust.

The 1978 *Cyana* dives revealed unusual lava formations and mineral deposits. Solidified lava lakes, probably formed by rapid outpourings of lava, were observed. Some of the lakes are hundreds of meters long and more than five meters deep. In places the surface of a lava lake has caved in, forming a collapse pit. Pillars and walls of basalt at the edge of the lakes are marked by bands of rapidly cooled basaltic glass that may record changes in the level of the lava. The bands were probably formed either during lateral outflows from the lava lakes or during the drainage of lava back into the underlying magma chamber.

The Hydrothermal Field

Near the boundary of zones 1 and 2 a chain of mounds several meters high was discovered. The geochemists in the group, led by Roger Hekinian, found that the mounds consist of sulfides of zinc, iron and copper with a small admixture of silver. It was suspected that the sulfide mounds were created by hydrothermal venting of fluids through the sea floor. There were three other indications that hydrothermal activity might be important in the area. In 1974 and again in 1977 temperature anomalies of several hundredths of a degree Celsius had been detected, and unusually high concentrations of helium 3 had been measured over the spreading center. This light isotope of helium is generally considered to be a reliable indicator of hydrothermal activity. In addition the workers making one of the *Cyana* dives observed large clamshells, similar to shells seen at the Galápagos hydrothermal vents, although none of the shells at the more northerly site were occupied by live clams. (As it happened the *Cyana* on this dive came within a few hundred meters of the vents discovered a year later by the *Alvin*.)

In 1979, shortly before the *Alvin* was brought on the scene, we made a brief photographic and mapping survey of the axis of the spreading center to the southwest of the *Cyana* dive site. We wanted to investigate the change in the geologic structures along the spreading center and to follow up on the tantalizing signs of hydrothermal activity. The time was well spent. Under the direction of Fred N. Spiess of the Scripps Institution, the Deep-Towed Instrument Package was deployed to obtain topographic and side-scanning-sonar profiles, extending our bathymetric chart to the

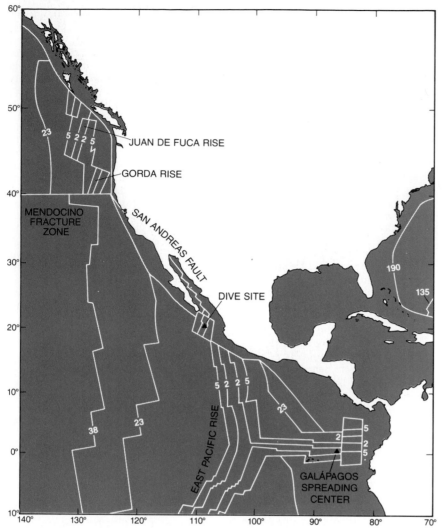

EAST PACIFIC RISE forms the boundary between the Pacific Plate and the Rivera Plate (a fragment of the North American Plate) in the region off the west coast of Mexico that was selected for exploration. The triangles mark the dive sites of the East Pacific Rise project and of the earlier expedition to the Galápagos spreading center off Peru. At both sites the oceanic crust is spreading at a rate of approximately six centimeters per year. Numbered contours indicate the age of the oceanic crust in millions of years. The map is based on one prepared by W. Pitman, R. Larson and E. Herron and on a chart published by the Geological Society of America.

southwest and delineating the axis of the spreading center.

The *Angus,* a strongly built camera sled that also carried a temperature sensor, was lowered to the sea floor for a series of long traverses a few meters above the rugged volcanic terrain. Guided by Robert D. Ballard of the Woods Hole Oceanographic Institution, the *Angus* detected elevated temperatures in several places and telemetered the information to the control ship on the surface. We immediately raised the camera and eagerly awaited the development of the film. Then we quickly spun through the roll of developed film until we found about a dozen frames that showed an assortment of bottom-dwelling creatures similar to those discovered two years earlier at the Galápagos spreading center. It appeared that the Galápagos hydrothermal vents and their associated biological communities were not unique.

The evidence was striking enough for us to shift the planned diving program to the southwest. The dive team was augmented by another dozen geologists and geophysicists from the U.S., France and Mexico, including the two of us. First a triangular array of ocean-bottom seismometers was precisely positioned in Zone 1 with the aid of signals relayed by acoustic transponders on the ocean floor. Several preliminary dives made with the *Alvin* demonstrated that seismic and gravitational measurements could be made successfully. On the third dive the hydrothermal vents were seen for the first time by Francheteau and one of us (Luyendyk).

It is difficult to convey the strange quality of such an experience. First one spends two hours or so in almost to-

tal darkness, dropping by gravity more than two and a half kilometers to the sea floor. Three people are huddled in the cold and cramped confines of the *Alvin*'s pressurized spherical compartment, which is only two meters in diameter. On approaching the bottom the submersible's running lights are turned on, and the illuminated water takes on a dim greenish glow. Minutes later the sea floor is sighted. As soon as the *Alvin* has reached the bottom, the team reports its position to the control ship and is given a course to steer to a bottom target. Edging ahead slowly (at about half a kilometer per hour) over the glistening volcanic rock, the investigators peer through the portholes, seeing only 10 or 15 meters into the darkness.

A New Ecosystem

On the dive in question we were making gravity measurements in the volcanic zone when we came on the hydrothermal field. The scene was like one out of an old horror movie. Shimmering water rose between the basaltic pillows along the axis of the neovolcanic zone. Large white clams as much as 30 centimeters long nestled between the black pillows; white crabs scampered blindly across the volcanic terrain. Most dramatic of all were the clusters of giant tube worms, some of them as long as three meters. These weird creatures appeared to live in dense colonies surrounding the vents, in water ranging in temperature from two to 20 degrees C. The worms, known as vestimentiferan pogonophorans, waved eerily in the hydrothermal currents, their bright red plumes extending well beyond their white protective tubes. (The red color of both the tubeworm plumes and the clam tissues results from the presence of oxygenated hemoglobin in their blood.) Occasionally a crab would climb the stalk of a tube worm, presumably to attack its plume.

A subsequent *Alvin* dive was guided directly to another hydrothermal area identified by the *Angus,* southwest of the first vents we visited. The sight here was even more dramatic: extremely hot fluids, blackened by sulfide precipitates, were blasting upward through chimney-like vents as much as 10 meters tall and 40 centimeters wide. We named the vents "black smokers." The chimneys protruded in clusters from mounds of sulfide precipitates. A puzzling structure seen from the *Cyana* the year before was probably a fossilized black smoker of this kind.

Our initial attempts to measure the temperature of the black fluids were unsuccessful. Until then the highest temperature recorded on the ocean floor was 21 degrees C., measured only two months earlier on the Galápagos spreading center. Our thermometer was calibrated to 32 degrees C.; when it was

UNUSUAL LIFE FORMS near the hydrothermal vents on the East Pacific Rise include many of the same organisms found for the first time two years earlier on an expedition to the Galápagos spreading center. The photograph, made from the *Alvin*, shows a cluster of giant tube worms waving in cooler (20 degrees C.) hydrothermal currents, where the water is not blackened by minerals. The worms, classified as vestimentiferan pogonophorans, are distinguished by bright red plumes extending from white protective sheaths; some of the tube worms are as long as three meters. Other organisms include clams and white crabs. At the base of the food chain are chemosynthetic bacteria; the community is independent of solar energy. The photograph was made by Fred N. Speiss of the Scripps Institution.

PILLOW LAVA covers much of the sea floor near the hydrothermal vents. The pillow shapes were formed by the rapid cooling of magma extruded through cracks in the crust during a volcanic eruption. In the background is an ocean-bottom seismometer; an array of such instruments was deployed from a surface ship. The photograph was made by John A. Orcutt of the Scripps Institution of Oceanography.

inserted into the first chimney, the reading immediately sailed off the scale. Moreover, when the probe was withdrawn, the plastic rod on which it was mounted showed signs of melting! The temperature probe was hastily recalibrated at sea and measurements were made on several more dives; they indicated temperatures of at least 350 degrees C., an estimate that was more precisely documented later by another dive team equipped with a thermometer modified for such a temperature range. The 350-degree water does not boil because the pressure at the depth of the vents is roughly 275 times atmospheric pressure.

The hydrothermal vents show considerable variety along the crest of the rise. To the northeast the vent waters are relatively clear; they are also cooler (less

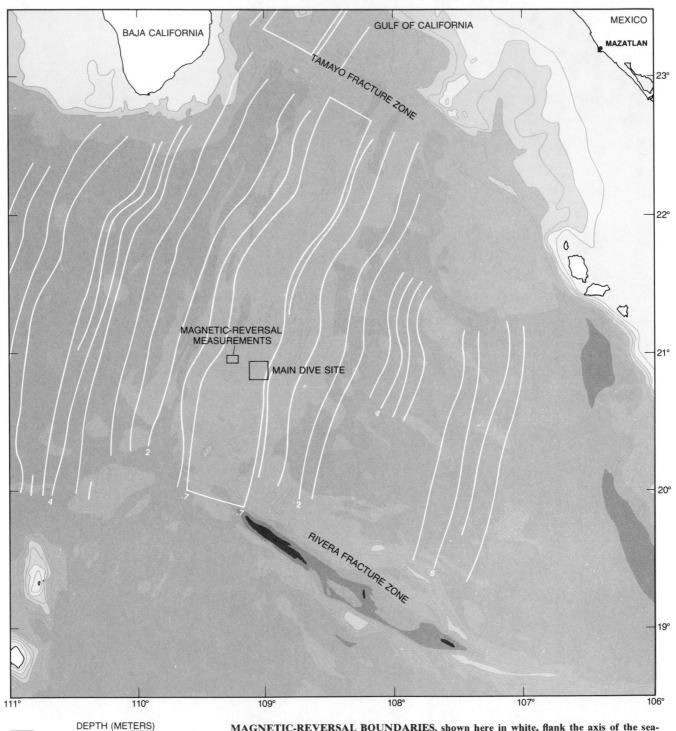

MAGNETIC-REVERSAL BOUNDARIES, shown here in white, flank the axis of the seafloor spreading center in the general area of the dive site on the East Pacific Rise. The white numbers at the ends of the lines give the age of the crust in millions of years. The rectangular area outlined in white is the central, positive-polarity magnetic stripe in which the rock is magnetized in the same direction as the earth's present magnetic field; it extends outward on each side of the axis to a distance corresponding to an age of 700,000 years. The outlying white lines parallel to the central stripe indicate boundaries where the remnant magnetism of the crustal rocks changes direction; hence the boundaries record reversals of the geomagnetic field. Colored contours, which are identified in the key at the left, give the depth of the ocean in meters.

than 20 degrees C.), and they diffuse slowly through the rocks. The densest biological communities are found here. Toward the southwest the vents emit hotter fluids laden with mineral precipitates. The rate of flow also increases to the southwest, culminating in the spectacular black-smoker vents. This rather regular gradation suggests the possibility of cycles in the intensity of volcanic and hydrothermal activity along the axis.

The biological community we had encountered turned out to be very similar to the one found in 1977 on the Galápagos spreading center. Certain distinctive brown mussels of the Galápagos site were absent, but sea anemones, worms called serpulids, crabs of the galatheid and brachyuran groups, large clams and giant tube worms all appeared to be the same. Each colony occupied an area roughly 30 meters wide and 100 meters long. The animals are attracted not by the warmth of the hydrothermal fluids but by the concentrated food supply. Nutrients are hundreds of times more plentiful near the vents than they are in the surrounding waters.

The food chain of this extraordinary ecological system has been analyzed by Robert R. Hessler of the Scripps Institution, J. Frederick Grassle of the Woods Hole Oceanographic Institution and others. At the base of the food chain are chemosynthetic bacteria that oxidize hydrogen sulfide emitted by the vents to form elemental sulfur and various sulfates. The bacteria harness the energy liberated by the oxidation in order to incorporate carbon dioxide into organic matter. Most of the larger organisms either feed on the bacteria by filtering or live with them symbiotically. Some are scavengers or predators. The communities are entirely independent of photosynthesis and indeed of solar energy; they rely instead on the flow of energy from the earth's interior. The fact that communities of this kind have been found both at the Galápagos spreading center and on the crest of the East Pacific Rise some 3,000 kilometers away suggests they may be distributed along much of the worldwide rift system. The communities must lead precarious lives as the hydrothermal vent systems turn on and off with sporadic volcanic cycles. Indeed, isolated piles of empty clamshells of uniform size bear witness to local mass fatalities.

Geochemical Implications

The discovery of hydrothermal vents on the axes of two Pacific spreading centers has revolutionized theories of the chemical budget of the oceans. Formerly it was held that seawater maintains an equilibrium between input processes (mainly the efflux of rivers) and output processes (the deposition of sediments and low-temperature chemical reactions between the seawater and the ocean floor). As more became known about mineral abundances and about low-temperature reactions of seawater with sedimentary and volcanic rocks, problems arose with the "bookkeeping" for certain elements. For example, more magnesium ions and sulfate ions are supplied by rivers than can be removed from the sea by sedimentation, by the formation of clays and by the weathering of basalt. In addition the ocean floor appears to accumulate far more manganese than rivers can supply.

The hydrothermal circulation of sea-

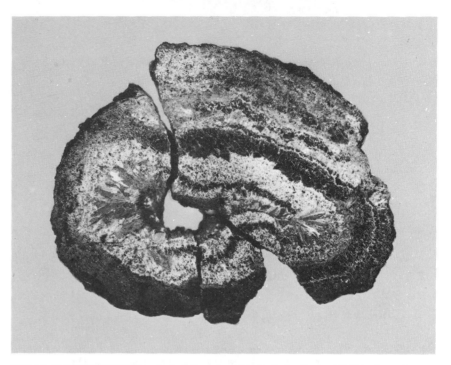

SMALL CHIMNEY VENT typical of those found in the hydrothermal field is shown in cross section. The concentric bands of minerals are chiefly sphalerite, pyrite and chalcopyrite, which are sulfides respectively of zinc, iron, and copper and iron. Transitions from one mineral to another reflect changes in the properties of the fluids emitted. The photograph was made by Rachel Haymon of the Scripps Institution, who also analyzed the mineral content of the chimney.

FILTERED PRECIPITATE from the effluent of a black-smoker chimney is shown in a scanning electron micrograph. The hexagonal platelets are crystals of pyrrhotite, a form of iron sulfide. The minerals recovered from the black vent water also include pyrite, sphalerite and other sulfides. The micrograph was made by J. Douglas Macdougall of the Scripps Institution.

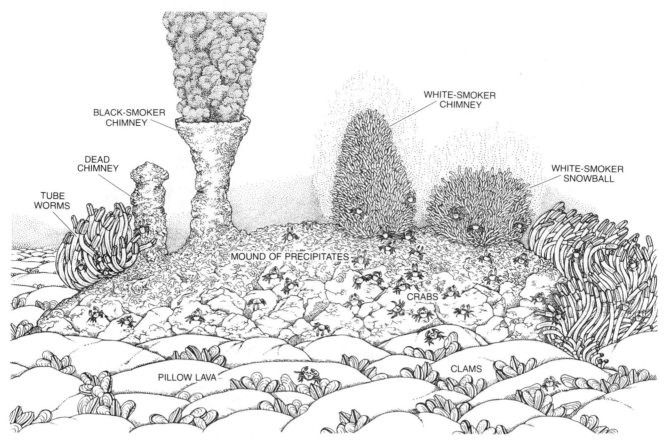

BLACK-SMOKER
CHIMNEY

DEAD
CHIMNEY

TUBE
WORMS

WHITE-SMOKER
CHIMNEY

WHITE-SMOKER
SNOWBALL

MOUND OF PRECIPITATES

CRABS

PILLOW LAVA

CLAMS

IDEALIZED SCENE in the hydrothermal field near the high-temperature smoker vents shows an array of typical vent structures on top of a mound of precipitated minerals and organic debris. The white-smoker chimney is built up out of burrows made by a little-understood organism called the Pompeii worm. The white, cloudy fluids emitted by the white smoker have temperatures of up to 300 degrees C. The hottest water, having temperatures of up to 350 degrees C., comes from the black smokers, whose chimneys are made up of sulfide precipitates. The mound rises from a terrain dominated by glistening black lava pillows; in the crevices between pillows clams live.

water along submarine rift systems introduces a new factor: high-temperature chemical exchanges between fluids and solids. According to John M. Edmond of the Massachusetts Institute of Technology, reactions between hot seawater and basaltic rock can convert dissolved sulfates into solid sulfate and sulfide minerals; similar reactions can remove magnesium ions and hydroxyl (–OH) ions from seawater and sequester them in hydrothermal clays. The hot seawater is converted by these reactions into a chemically reduced, acidic solution that leaches calcium, silicon, manganese, iron, lithium and other positively charged ions from the rocks and releases them into the ocean. In this way hydrothermal systems can balance the budget for the major mineral constituents of seawater; the hydrothermal circulation

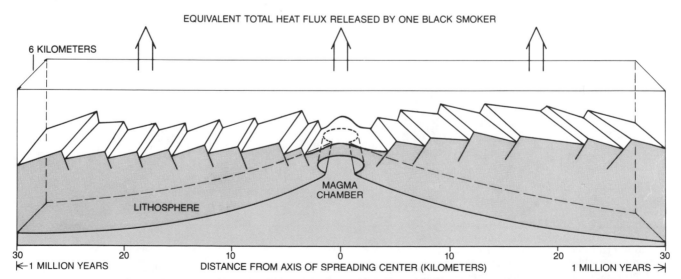

EQUIVALENT TOTAL HEAT FLUX RELEASED BY ONE BLACK SMOKER

6 KILOMETERS

MAGMA
CHAMBER

LITHOSPHERE

30   20   10   0   10   20   30
←1 MILLION YEARS   DISTANCE FROM AXIS OF SPREADING CENTER (KILOMETERS)   1 MILLION YEARS →

HEAT OUTPUT from the hydrothermal vents makes a significant contribution to the total heat budget of the earth. The heat output of a single black-smoker vent is compared here with the heat transmitted through the crust by conduction. The output of the vent is equivalent to the conductive flux of a segment of the rift system measuring some six kilometers along the axis and 60 kilometers across it.

can also account for the observed concentration and distribution of numerous minor and trace constituents.

Edmond has found that the hydrothermal effluents of the Galápagos system mix with normal seawater during their ascent through the volcanic rock. The mixing lowers the temperature of the fluids and leads to the deposition of minerals within the rocks, thereby altering the chemistry of the hot springs that discharge onto the sea floor. The high temperature and the chemical composition of the springs at the East Pacific Rise site indicate that here the hydrothermal fluids do not mix significantly with cool seawater on their way to the sea floor. Hence the fluids represent the true hydrothermal contribution to the marine chemical cycle. The water issuing from the vents has descended most of the way to the magma chamber before returning to the sea floor.

The undiluted hydrothermal fluids at the East Pacific Rise site apparently become blackened with fine-grained precipitates of iron sulfide and zinc sulfide as contact is made with cold, alkaline seawater at the ocean floor. Preliminary analyses by Rachel Haymon and Miriam Kastner of the Scripps Institution indicate that the mounds and chimneys around the vents are composed mostly of sulfides of zinc, iron and copper and sulfates of calcium and magnesium. The precise mechanisms by which the minerals are formed, the rate at which they are deposited and the ratio of water to rock at various points in the system are currently subjects of debate among geochemists. There is no doubt, however, that the vents will have a central role in models of the chemistry of the oceans.

Geophysical Experiments

Most of our geophysics program was directed toward learning more about the relation of the conjectured axial magma chamber to tectonic, volcanic and hydrothermal activity on the sea floor. We were fortunate in being able to do the experiments in an area where hydrothermal activity was proceeding at such a brisk pace. Measurements of seismic velocity, earthquake activity, gravitational anomalies, electrical conductivity and magnetic polarity were made on the crest of the rise in order to probe the subsurface structure there.

For geophysicists as well as for geochemists and biologists the most impressive finding of the expedition was the field of active hydrothermal vents. Such hydrothermal circulation on mid-ocean ridges had been proposed on theoretical grounds 15 years earlier, but it had proved difficult to study.

It is possible to calculate a theoretical rate at which newly formed lithosphere cools by means of heat conduction. The calculation in turn specifies an expected rate of conductive heat flow through the mid-ocean ridges. Measurements of purely conductive heat flow near the crest of the rise, however, yield values almost an order of magnitude smaller than the theoretical models. Are the models wrong, or is the circulation of near-freezing seawater in the hot, newly formed crust cooling the crust through convective heat transfer at a much higher rate? How deep does the seawater penetrate into the oceanic crust and how wide is the discharge zone? How is the chemical composition of the crust affected by the circulation and what minerals are deposited? A key geophysical consideration in all these questions is the depth of the cracks and fissures in the rocks at the spreading center. Bounds on the depth can be determined by measurements of seismic velocity, electrical conductivity and gravitational anomalies.

The large discrepancy between the measured conductive heat flow through the sea floor and the value predicted by models of the cooling lithosphere suggests that at least a third of the heat loss at the mid-ocean ridges is brought about by nonconductive means and presumably by hydrothermal circulation. At the Galápagos spreading center, where hydrothermal activity was observed for the first time, estimates of the heat flow were hindered by the diffuse nature of the circulation. The conditions at the East Pacific Rise site are more favorable for such measurements.

From careful viewing of motion-picture films and videotapes of the vents, flow rates were estimated. Given an average exit velocity of two or three meters per second, a single vent accounts for a heat flow of some 60 million calories per second. That is between three and six times the total theoretical heat loss across a kilometer-long segment of the mid-ocean ridge out to a distance of 30 kilometers on each side. At least 12 major chimneys were found in the southwestern part of the study area, and so the total heat flux is evidently very large. Indeed, the heat loss is so large that it suggests the lifetime of a vent is probably short. The lifetime may be only several years.

The vents appear to be confined to a narrow, linear region a few hundred meters wide and six kilometers long within the neovolcanic zone. In this band 12 temperature anomalies were detected and verified photographically, and eight vents were visited and studied with the

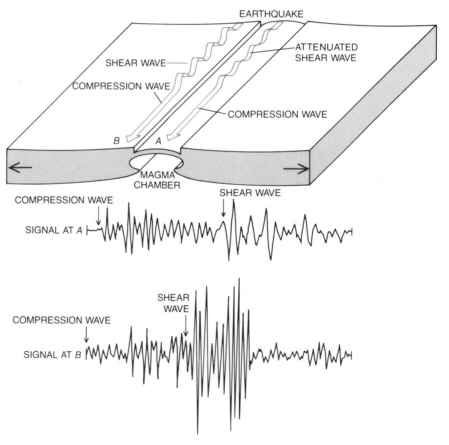

SEISMIC SHEAR WAVES, or vibrations that oscillate perpendicular to the direction of propagation, are attenuated by passage through the magma chamber that is thought to lie under the axis of the spreading center. When the seismic signal from a distant earthquake is recorded by an ocean-bottom seismometer on the axis (*point A*), compressional waves are strongly received but shear waves are much reduced in amplitude. On a parallel path only 10 kilometers from the axis (*point B*) both compressional waves and shear waves are transmitted efficiently. This finding suggests that a magma chamber exists under the axis and that it is relatively narrow.

Alvin. In general the cooler vents toward the northeast are surrounded by Galápagos-style biological communities. The hottest vents near the southwestern end also have unusual life forms associated with them, but the animals tend to live at a safe distance (several meters) from the vents.

Seismic Measurements

A high-resolution seismic experiment was designed by John A. Orcutt and one of us (Macdonald) to determine the depth of the cracking and fissuring in the rocks along the crest of the rise. We proposed to measure the velocity of seismic waves in the upper crust as a function of depth. Because the spherical spreading of energy from an explosive charge detonated at the surface of the sea leads to reverberations from nearby topographic features, the resolution of seismic measurements in the top 1,000 meters of the crust is degraded. In order to overcome this difficulty it was necessary to devise a method of placing both the sources and the receivers of the seismic waves on the sea floor. For the experiment to work we also needed to time events to within a millisecond or so.

The *Alvin* provided a way of solving both problems. Explosive charges are virtually useless as seismic sources at the high pressure encountered at great depth, and so we attached a hydraulic hammer to the *Alvin* to serve as a source of seismic waves. To establish a precise time base the *Alvin* was driven up to within two meters of each of the ocean-bottom seismometers for a calibration "thump" by the hammer on the sea floor; the response was recorded by a sensor on board the *Alvin* as well as by the seismometer. The submersible returned to recalibrate each seismometer at the end of the dive.

In four dives we completed a seismic-refraction profile 1,000 meters long parallel to the spreading center and a profile 800 meters long across the spreading center. The analysis of the data is not yet completed, but simple travel-time calculations yield a preliminary determination of the velocity of seismic waves at the surface of the crust parallel to the axis of the spreading center. We obtained a seismic velocity of 3.3 kilometers per second, which is quite slow compared with the laboratory value for basalt at the same pressure (approximately 5.5 kilometers per second). Evidently the reason for the low velocity is the pervasive cracking and porosity of the rock. No major faults or fissures were traversed, but numerous hairline cracks and voids were observed in the lava pillows. Detailed conclusions concerning the degree of cracking and porosity needed to account for the observations will have to await measurements of the physical properties of the rock samples and the completion of seismic analyses at longer ranges. Of particular interest will be the depth at which the seismic velocity rises above five kilometers per second, indicating the closure of most of the fissures.

The results of an earlier seismic-refraction experiment had indicated the presence of a magma chamber at a depth of from two to three kilometers under the hydrothermal field. In the earlier experiment, which was less precise but of larger scale, explosive charges were detonated from a surface ship at ranges of up to 60 kilometers from a triangular array of ocean-bottom seismometers. At a depth of only two kilometers under the sea floor a low-velocity zone for compressional seismic waves was observed, indicating the presence of partially melted rock. At points 10 kilometers off the axis of the spreading center the seismic velocity was found to be normal for oceanic basalt or somewhat higher than normal. The magma chamber therefore appears to be confined to a zone 20 kilometers wide centered on the axis of the spreading zone.

Earthquakes and Volcanism

A second item of evidence for the existence of an axial magma chamber comes from measurements of the propagation of seismic shear waves generated by earthquakes. Where rock is partially melted, as it would be in a magma chamber, shear waves are strongly damped. Orcutt and two of his colleagues, Ian Reid and William A. Prothero, Jr., found that for paths along the spreading center shear waves from

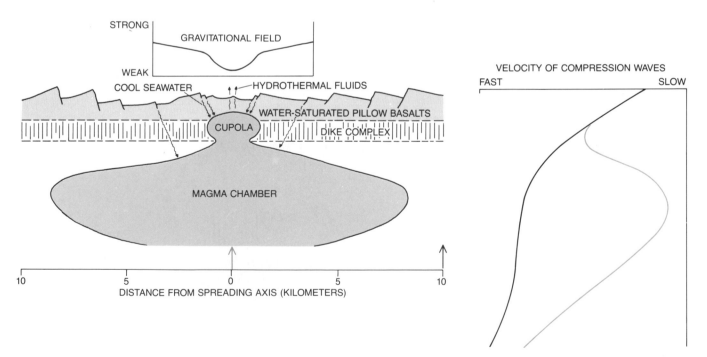

MODEL OF THE EARTH'S CRUST under the East Pacific Rise is represented by a cross-sectional drawing based mainly on geophysical measurements. Cool seawater percolates through cracks and fissures in the rocks of the spreading center, approaching the molten rock in an elongated cupola above the main magma chamber. The water is heated by the magma and is expelled through vents along the axis of the spreading center. The existence of a cupola and of hydrothermal activity is thought to be episodic, generally following periods of intense volcanism and tectonic motion. The geophysical properties graphed at the right are sensitive both to the permeability of the crust

earthquakes are severely attenuated, whereas only 10 kilometers from the axis shear waves are transmitted efficiently. Again this finding is clear (although indirect) evidence for the presence of a shallow, narrow magma chamber under the spreading center.

Basalt samples collected with the *Alvin* have been analyzed by James W. Hawkins of the Scripps Institution; the nature of the rocks also suggests the presence of a shallow axial magma chamber. The basalt samples, collected along a six-kilometer traverse of the neovolcanic zone, were found to have a limited range of compositions, indicating they were derived from a single parent magma by fractional crystallization of the minerals olivine and plagioclase at comparatively low pressure. The results are consistent with a magma chamber less than six kilometers deep, which in turn implies that the roof of the chamber is only a few kilometers thick. This thin covering layer is densely cracked and fissured, allowing seawater to percolate deep enough into the crust to be heated to at least 350 degrees C.

How deep does the circulation extend? As a sequel to the ocean-bottom seismic measurements we returned last summer to measure microearthquakes in the hydrothermal field. Seven ocean-bottom seismometers were guided into position with the aid of moored acoustic transponders that had been left to mark the vents. If the hydrothermal flow had some telltale seismic "signature," perhaps we could determine how deep the flow extends. So far the results have been encouraging. Earthquakes in the region appear to be shallow: only two or three kilometers deep at the most. This finding is in excellent agreement with earlier indications that the roof of the magma chamber is thin, and it establishes a reasonable upper limit for the depth of the cracked part of the crust.

Among the seismic events recorded by the array were small earthquakes of a distinctive kind called harmonic tremors. Such tremors were associated with the catastrophic eruption of Mount St. Helens a year ago, and they are a sure sign of impending or recent volcanic activity. Just before and during the Mount St. Helens eruption the tremors increased in frequency until they became almost continuous. Our records for the East Pacific Rise look similar, showing as many as several hundred events per hour. It is possible that this segment of the rise is now either ending or entering an active volcanic stage.

Gravitational Anomalies

Closely associated with the seismic experiments was a series of gravity measurements conducted with the *Alvin* by Spiess and one of us (Luyendyk). As in the case of seismic waves, the local gravitational field should be altered by variations in the density of the crust caused by fissuring or by the presence of a shallow magma chamber. The expected gravitational anomalies are small, and they are difficult to measure from the ocean surface because of the considerable distance of the sensor from the source and because of spurious acceleration signals recorded by shipboard sensors. Once again the *Alvin* provided a solution. Measuring the gravitational field from the *Alvin* while sitting quietly on the sea floor had the effect of reducing the spurious accelerations. Moreover, making the measurements closer to the source enhanced the signal recorded by the gravimeter.

The gravitation measurements, which were made along a seven-kilometer profile extending from Zone 1 to Zone 3, showed a pronounced negative gravity anomaly over the neovolcanic zone. The anomaly is apparently centered on the central volcanic ridge and occupies much of zones 1 and 2. It indicates a region of lower-than-average density, which could be caused either by fissuring of the crust or by a shallow magma chamber. Geologic observations show a maximum of fissuring in Zone 2, whereas the negative gravity anomaly is centered on Zone 1, which has comparatively few fissures. This observation suggests (but does not fully demonstrate) that the gravity anomaly is caused by a shallow magma chamber.

If one assumes that the magma chamber has the shape of a horizontal cylinder with its axis parallel to the ridge and its center directly under the measured minimum in the gravitational field, the gravitational data indicate that the center of the cylinder is about 1,000 meters below the sea floor. If one further assumes that the chamber is filled with molten basalt, its density would have to be lower than the density of the surrounding rock by about .21 gram per cubic centimeter. This estimate in turn implies that the upper edge of the cylinder is about 600 meters below the sea floor. Assuming a smaller density contrast between the magma in the chamber and the surrounding rock would require that the magma body be larger and reach closer to the surface of the crust.

The seismic results make it seem more likely that the main magma chamber is much larger and deeper. Indeed, for geologic reasons it seems probable that the main, permanent part of the chamber is between two and six kilometers below the sea floor and is two or three times as wide as it is deep. The structure apparent in the gravity data may be not the main chamber but a small, transitory, elongated cupola, or dome, at the apex of the main chamber. Such a cupola could occupy all of Zone 1 and could supply the magma to feed the lava flows found there.

The magnitude of the negative gravity anomaly corresponds to a mass deficiency under the axis of the spreading center

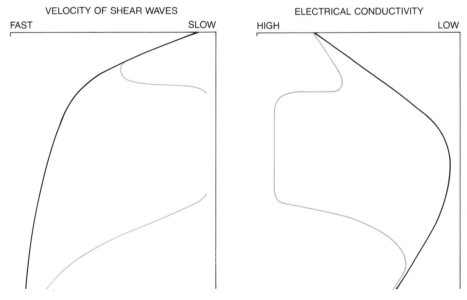

VELOCITY OF SHEAR WAVES

FAST   SLOW

ELECTRICAL CONDUCTIVITY

HIGH   LOW

and to the presence of the axial magma chamber. Colored curves show the variation of each property with depth at the spreading center; black curves show variation with depth approximately 10 kilometers from the spreading center. A measured gravitational anomaly is also shown schematically; such variations in the strength of the gravitational field, which were detected at the sea floor, reveal the shallow structure of the crust and of the magma chamber.

of about 90 million kilograms per meter of ridge. If the crust at the spreading center is in isostatic equilibrium (meaning that the forces of gravitation and buoyancy are in balance), a mass excess should exist somewhere on the ocean bottom to balance the deeper mass deficiency. Zone 1 is an uplifted topographic block about a kilometer wide and roughly 20 to 30 meters higher than the surrounding terrain. Even so, it is only about half as high as it would have to be to achieve isostatic equilibrium at the axis of the spreading center. Either the mass deficiency is balanced by other topographic features farther away from the axis, or friction along fault planes is holding down the central block in opposition to the buoyant force.

Electrical Measurements

The measurements of gravitational anomalies also yield an estimate of the bulk density of the upper 100 meters of the bottom topography. Along the crest of the rise the bulk density is about 2.6 grams per cubic centimeter, compared with a measurement of 2.9 grams per cubic centimeter for some 90 rock samples taken from the bottom. The discrepancy suggests that the topography has a porosity of about 15 percent.

The presence of water in the rocks of the crust might be detected directly by means of the electrical conductivity of water. At greater depths electrical conductivity might also be exploited to detect the axial magma chamber, because magma has a much higher conductivity than solid basalt. For these purposes a new electrical-sounding technique was developed by Charles S. Cox of the Scripps Institution. The experiment was designed to give further information on the percolation of seawater into the crust, on the depth of fissuring and on the lateral extent of the magma chamber. No similar measurements had been attempted before, nor had any other techniques been able to estimate the conductivity of the crust under the oceans.

An electric-dipole antenna 800 meters long was towed near the sea floor behind the research vessel *Melville*, which is operated by the Scripps Institution. The antenna transmitted electrical signals into the ocean and the crust at frequencies selected so that the signal would be quickly absorbed in the ocean, whereas it might penetrate to a considerable depth in the crust. Three receivers were placed on the sea floor near the spreading center. Because the experiment was our first attempt at towing a large, frag-

ile antenna near the sea floor the rugged topography of Zone 1 was avoided. The electrical soundings were made in an area between 10 and 15 kilometers west of the spreading axis in crust that is between 300,000 and 400,000 years old.

Soundings were made to a depth of approximately eight kilometers below the sea floor. The pattern of conductivity indicates that only 10 to 15 kilometers from the spreading axis the magma chamber is less than 200 meters thick. This finding is a strong confirmation of the seismic experiments that had suggested a narrow axial magma chamber. Observations made with the *Cyana* indicate that active faulting of the crust diminishes at a distance of between 10 and 12 kilometers from the axis. Perhaps the width of the magma chamber controls the width of active tectonic faulting on moderate-to-fast spreading centers. Another finding of the electrical-sounding experiment is that conductivity comparatively near the surface of the sea bed is low. This indicates that seawater penetrates the crust to a maximum depth of from two to four kilometers.

Magnetic Reversals

In another series of dives we carried our investigation slightly off the axis of the ridge and northwest to the edge of the youngest major magnetic-reversal stripe. The object was to see what the geometry of such a boundary could tell us about the formation of new oceanic crust along the spreading axis. Almost two decades after the Vine-Matthews model was proposed there is still no full understanding of how such stripes are formed. At one time it was widely believed most of the magnetic signal associated with the magnetic-reversal stripes was confined to the top 500 meters of the oceanic crust. Deep-sea drilling into the deeper layers of the crust has revealed a chaotic mixture of magnetic polarities and intensities that is quite out of character with the linear sequence of magnetic stripes measured from the surface of the sea.

In the first stage of this study we constructed a three-dimensional mathematical model of the magnetic-reversal boundary on the basis of earlier measurements made with a deep-towed magnetometer. From the model we calculated that the boundary was remarkably straight and narrow: less than 1.4 kilometers wide. Filtering of the data was required in order to achieve a stable solution, however, and it made us wonder if the boundary was really so well defined. To find out how the oppositely magnetized stripes are actually arrayed on the sea floor we mounted a more sensitive magnetometer on the *Alvin*. The instrument measured the components of the magnetic field in three dimensions

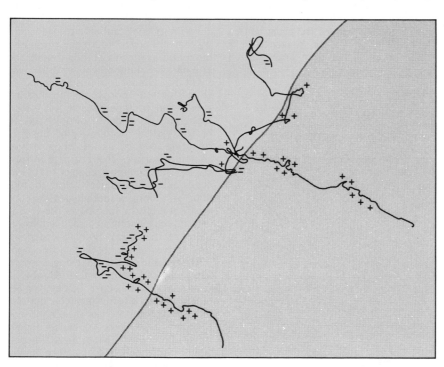

MAGNETIC POLARITY OF THE CRUST was measured with instruments mounted on the *Alvin* at more than 250 points along a series of traverses near a magnetic-reversal boundary. Plus signs designate positive polarity (identical with the present direction of the field) and minus signs negative polarity. The measurements made on the sea floor with the *Alvin* are superposed on a map of the magnetic-reversal boundary based on data recorded from a surface ship. On this map the gray area is positive and the colored area is negative. The transition zone in each case is remarkably sharp, indicating that the oceanic crust is created in a narrow band (about a kilometer wide). The boundary suggested by the ocean-bottom measurements is about 500 meters northwest of the position calculated from the data recorded at the surface, possibly because of the spillover of positively polarized material on top of negatively polarized rock.

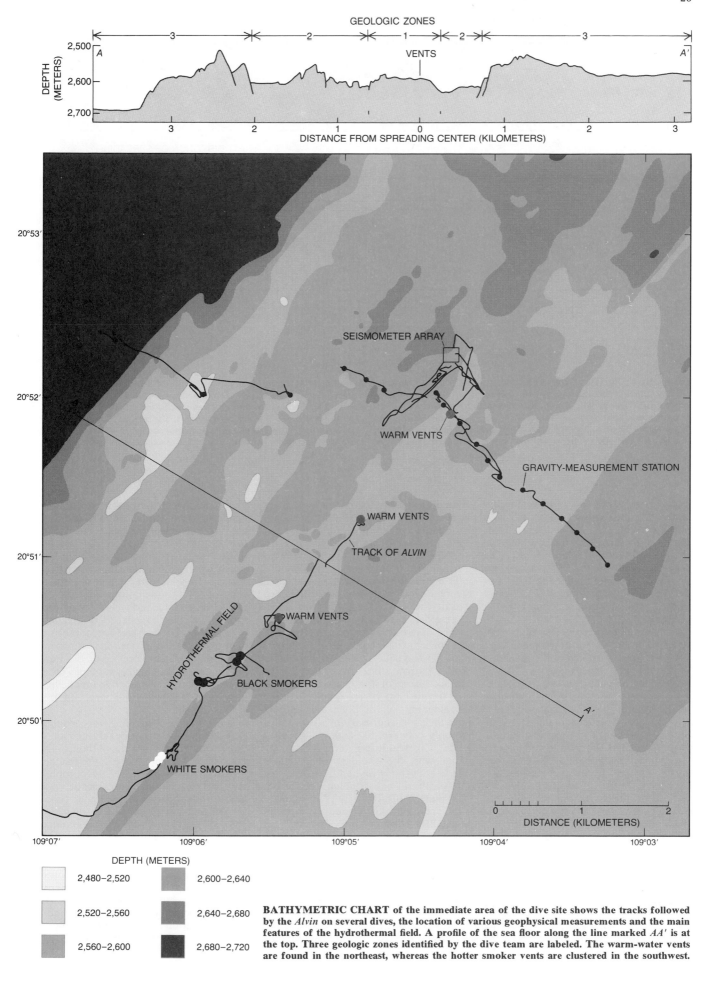

GEOLOGIC ZONES

BATHYMETRIC CHART of the immediate area of the dive site shows the tracks followed by the *Alvin* on several dives, the location of various geophysical measurements and the main features of the hydrothermal field. A profile of the sea floor along the line marked *AA'* is at the top. Three geologic zones identified by the dive team are labeled. The warm-water vents are found in the northeast, whereas the hotter smoker vents are clustered in the southwest.

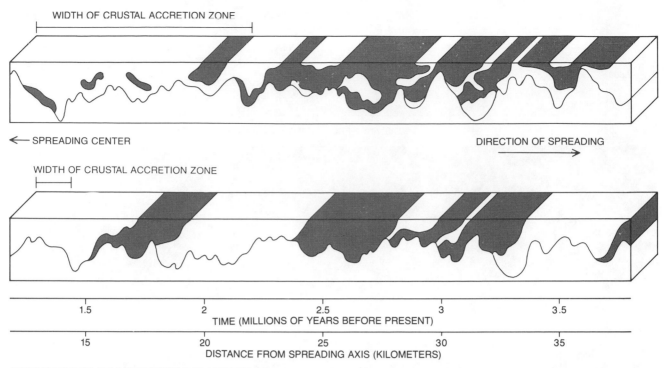

WIDTH OF CRUSTAL ACCRETION ZONE

← SPREADING CENTER   DIRECTION OF SPREADING

WIDTH OF CRUSTAL ACCRETION ZONE

| 1.5 | 2 | 2.5 | 3 | 3.5 |

TIME (MILLIONS OF YEARS BEFORE PRESENT)

| 15 | 20 | 25 | 30 | 35 |

DISTANCE FROM SPREADING AXIS (KILOMETERS)

DISTRIBUTION OF MAGNETIC POLARITIES in the oceanic crust is influenced by the width of the spreading zone, as is shown in two hypothetical profiles developed statistically by Hans Schouten and Charles Denham of the Woods Hole Oceanographic Institution. In the upper diagram the rate of sea-floor spreading is assumed to be low and the accretion zone, where new crust emerges, is assumed to be large. Such conditions prevail at the Mid-Atlantic Ridge. In the lower diagram the accretion zone is narrow, as it is at the East Pacific Rise. The thickness of the crust depends on the rate of spreading and on the frequency of volcanic eruptions. The degree of mixing of oppositely magnetized sections depends on the width of the accretion zone. The zone is thought to be wider at centers that spread slowly.

and also measured the vertical gradient in the field: the rate at which the field changes with height above the sea floor. In this first attempt to study a submarine magnetic-reversal boundary at close range we were joined by Loren Shure of the Scripps Institution and by Stephen P. Miller and Tanya M. Atwater of the University of California at Santa Barbara.

In five dives across the magnetic-reversal boundary we made more than 250 clear identifications of the magnetic polarity of the basalt outcrops. The results of the magnetic survey were striking. Even on long traverses on both sides of the boundary every magnetic target had the correct polarity, that is, the same polarity as that of the regional magnetic stripe defined by the deep-towed magnetometer. This observation was not too surprising for the younger side of the boundary, because newer positive-polarity crust (crust that has the same polarity as the earth's present magnetic field) would be expected to overlie older, negatively polarized crust. What was surprising was that there were no outlying regions of new crust on the older side of the boundary.

We found that the magnetic-reversal boundary surveyed with the *Alvin* is displaced about 500 meters farther northwest of the spreading axis than the boundary position calculated from the data collected by the deep-towed magnetometer. The calculated boundary marks the average position of the magnetic reversal in a cross section of crust to some depth. The fact that the boundary mapped directly on the sea floor is 500 meters northwest of the calculated average position indicates a spillover of basalt away from the volcanic vents over older, negatively polarized crust.

The data obtained with the *Alvin* and the earlier measurements together show that the magnetic stripes are formed in a very narrow zone. After allowing for the extension of the crust by faulting and for the finite period required for the geomagnetic field to reverse, it appears that the zone of crustal formation must be only between 500 and 1,000 meters wide. This result is in excellent agreement with the geologic observations made with the *Alvin* and the *Cyana* for the width of Zone 1, which varies from 400 meters to 1,200 meters at the present spreading axis. Hence the zone of crustal formation both today and about 700,000 years ago is sharply defined: it is barely a kilometer wide. Considering the lateral dimensions of the Pacific and North American plates, which are thousands of kilometers across, it is remarkable that the spreading center between them is so narrow and so stable.

How can this picture of well-ordered stripes be reconciled with the complex magnetic stratigraphy observed in the deep-sea drilling cores? It turns out that the holes penetrating deeper than 500 meters into the oceanic crust were all drilled in the Atlantic basin, which has much lower spreading rates. Statistical studies show that at such low rates a zone of crustal formation wider than a few kilometers generates a crustal section with a complex mixture of magnetic polarities. Given faster spreading and a narrower zone of crustal formation (the conditions that prevail at the East Pacific Rise), the magnetic stripes have a greater tendency toward magnetic homogeneity and the boundaries of the stripes are sharper.

It follows from this line of reasoning that the physical processes giving rise to new oceanic crust in the Atlantic are quite different from those in the Pacific. Further efforts to unravel the complex physical and chemical processes at work near the boundaries of the major lithospheric plates will require additional on-site experiments, in which the manned submersible will continue to have an essential part. One of the most rewarding outcomes of these studies is the unification of the many disparate fields within oceanography. Biologists, geochemists, geologists, geophysicists and physicists have all been drawn into a common quest for an understanding of sea-floor spreading at the mid-ocean ridges.

3

Hot Spots on The Earth's Surface

by Kevin C. Burke
and J. Tuzo Wilson
August 1976

These regions of unusual volcanic activity record the passage of plates over the face of the earth. They may also contribute to the fracturing of continents and the opening of new oceans

Scattered around the globe are more than 100 small regions of isolated volcanic activity known to geologists as hot spots. Unlike most of the world's volcanoes, they are not always found at the boundaries of the great drifting plates that make up the earth's surface; on the contrary, many of them lie deep in the interior of a plate. Most of the hot spots move only slowly, and in some cases the movement of the plates past them has left trails of extinct volcanoes. The hot spots and their volcanic trails are milestones that mark the passage of the plates.

That the plates are moving is now beyond dispute. Africa and South America, for example, are receding from each other as new material is injected into the sea floor between them. The complementary coastlines and certain geological features that seem to span the ocean are reminders of where the two continents were once joined. The relative motion of the plates carrying these continents has been reconstructed in detail, but the motion of one plate with respect to another cannot readily be translated into motion with respect to the earth's interior. It is not possible to determine whether both continents are moving (in opposite directions) or whether one continent is stationary and the other is drifting away from it. Hot spots, anchored in the deeper layers of the earth, provide the measuring instruments needed to resolve the question. From an analysis of the hot-spot population it appears that the African plate is stationary and that it has not moved during the past 30 million years.

The significance of hot spots is not confined to their role as a frame of reference. It now appears that they also have an important influence on the geophysical processes that propel the plates across the globe. When a continental plate comes to rest over a hot spot, the material welling up from deeper layers creates a broad dome. As the dome grows it develops deep fissures; in at least a few cases the continent may rupture entirely along some of these fissures, so that the hot spot initiates the formation of a new ocean. Thus just as earlier theories have explained the mobility of the continents, so hot spots may explain their mutability.

Plate Tectonics

The modern theory of plate tectonics divides the superficial regions of the earth into two layers. The lithosphere, the outermost layer and the only one directly accessible to us, is cold and rigid. Below it is the asthenosphere, which is white hot and capable of being slowly deformed. The asthenosphere is not liquid, although there is a small amount of melted rock in the earth's interior. The asthenosphere is a solid, but one that flows under stress. It is not unlike ice, which seems brittle in the form of an ice cube but is quite plastic in a glacier flowing down a mountain valley.

The distinction between lithosphere and asthenosphere is based on rigidity and to a large extent reflects differences in temperature. An older distinction, based on chemical composition, divides the upper earth into the crust and the mantle. The boundary between these layers does not correspond to that between the lithosphere and the asthenosphere. The crust is the upper portion of the lithosphere, and the lithosphere also contains the topmost part of the mantle. The asthenosphere usually lies entirely within the mantle.

Under the oceans the crust is composed primarily of basalt; the continents, on the other hand, are made largely of granitic rock. Granite is lighter than basalt, and the continents are considerably thicker than the oceanic crust, with the result that the continents float well above the ocean floor. It was once proposed that the continents move through the ocean floor like ships, but that hypothesis had to be abandoned. Actually the continents are carried by the lithosphere like rafts locked in the ice of a frozen river.

The lithosphere is broken into about a dozen plates, in which the continents are firmly anchored. The plates separate from one another at the crests of the mid-ocean ridges, where new lithosphere is created. The ridges wind through all the world's oceans and constitute the largest mountain system on the earth. At the crests of the ridges undersea volcanism adds new material to the plates, pushing them apart. The opposite process—the consumption of lithospheric plates—is observed where the plates converge and overlap. In those regions, called subduction zones, one plate plunges under another and is reabsorbed into the mantle.

The movement of the lithospheric plates is thought to be associated with large-scale convection currents in the mantle. The currents may actually drive the plate movements, but too little is known about convection in the mantle to warrant firm conclusions.

Hot Spots and Plumes

Almost all volcanic activity is confined to the margins of the plates. Along the full length of the mid-ocean ridges there is undersea volcanism in which the lava erupted is predominantly basalt. At convergent plate boundaries lavas are formed by the melting of lighter constituents of the subducted plate. The upwelling lava can create an island arc, such as the arcs of the Philippines, Japan and the Aleutians, or a volcanic mountain system, such as the Andes and the Cascade Range of the Americas. The lavas associated with convergent plates differ from the basalts of the mid-ocean ridges. They are called andesite lavas and they contain more silicon, calcium, sodium and potassium than basalt and less iron and magnesium.

Volcanism that is not associated with plate margins accounts for a small proportion of the world's volcanic activity, probably much less than 1 percent. It is these few isolated volcanoes that have been named hot spots. They are distin-

HOT SPOT in North Africa is an isolated group of volcanic mountains surrounded by the Sahara. The group is called Tibesti, and it lies in northeast Chad, near the Libyan border. The photograph was made from an altitude of about 920 kilometers by the LANDSAT earth-resources satellite. Dark blotches on the landscape are relatively recent lava flows. Two large, recent volcanic craters are visible; the one at the lower right is Emi Koussi, which has an elevation of 3,415 meters and is the highest peak in the region. Some older craters are recognizable, but they have been severely eroded. At Tibesti and at other African hot spots lavas of different ages are piled on top of one another, suggesting that the continent is stationary with respect to the hot spots and probably has been for 30 million years.

guished by their very isolation: in the middle of a rigid lithospheric plate, far from centers of seismic activity, a hot spot may be the only distinctive feature in an otherwise monotonous landscape. Almost all hot spots are regions of broad crustal uplift, and this swelling is distinct from the smaller-scale mountain-forming or island-forming activity characteristic of all volcanoes. Finally, the lavas associated with hot spots differ from those found both at the mid-ocean ridges and at subduction zones. The hot-spot lavas are basalts, like those of the ocean ridges, but they contain larger amounts of the alkali metals (lithium, sodium, potassium and so on). Alkali-rich lavas are rare at plate margins.

The mechanism that generates hot spots must be sought in the mantle. They may be surface manifestations of "plumes": rising, columnar currents of hot but solid material. The plumes might well up from below the asthenosphere, at a phase-change boundary a few hundred miles inside the mantle. The distinctive composition of the hot-spot lavas argues that their source is isolated from the general circulation pattern of the mantle. For example, plumes might be generated in stagnant regions in the center of a circular convection current, or they might come from a very deep layer of the mantle, below the region that is effectively stirred by convection. The circulation of the mantle is still poorly understood, however, and for the moment any attempt to explain the origin of hot spots must remain speculative. Here we are concerned mainly with surface manifestations, which are not strongly dependent on the exact source of the magma. It is possible to formulate a consistent interpretation of hot spots even without a detailed model of the earth's interior.

Island Chains

Perhaps the most prominent and most easily recognized hot spot is the one that has formed the Hawaiian Islands. On an expedition to the South Seas in 1838, James Dwight Dana, an American geologist, noted that these islands become progressively older as one proceeds northwest from Kilauea and Mauna Loa, the active volcanoes on Hawaii itself. (Dana estimated the ages of the islands from the extent to which they had been eroded.)

It is now apparent that all the islands in the Hawaiian chain were created by a single source of lava, over which the Pacific plate has passed on a course proceeding roughly toward the northwest. The plate has carried off a trail of volcanoes of increasing age, in much the same way that wind passing over a chimney carries off puffs of smoke.

Dana also called attention to two other chains of Pacific islands whose trend is parallel to that of the Hawaiian chain. These are the islands of the Austral Ridge and the Tuamotu Ridge; the latter group includes Pitcairn Island. Like the Hawaiian Islands, these chains become older toward the northwest, and in each of them the most recent volcanic activity is near the eastern terminus. It would be difficult to ignore the inference that all three chains were generated as a result of the same plate motion. Indeed, from the configuration of the islands the apparent course of the plate can be mapped.

Leonhard Euler, the 18th-century Swiss mathematician, proved that on the surface of a sphere the only possible motions are rotations. It is therefore always possible to describe the movement of a lithospheric plate as a rotation around a pole. (The pole does not have to pass through the plate itself.) W. Jason Morgan of Princeton University has been able to show that the Hawaiian, Austral and Tuamotu chains could all have been generated by the rotation of the Pacific plate around the same pole. Employing a somewhat different approach, Jean-Bernard Minster and his colleagues at the California Institute of Technology have deduced from observed rates of sea-floor spreading that if the African plate has been stationary, then the Pacific plate must have moved along the trajectory defined by the Hawaiian chain.

At the western end of the Hawaiian Islands a string of submerged mountains, the Emperor Seamounts, strikes to the north. It is appealing to consider the entire system of islands and seamounts as a single chain that has changed direction, and age determinations support that interpretation. The oldest of the Hawaiian Islands, near the bend, are about 40 million years old. The Emperor Seamounts continue the age sequence without interruption, beginning near the bend with an age of 40 million years and continuing to an age of about 80 million years where the chain ends off the Kamchatka Peninsula. Morgan has found he can account for the formation of the seamounts by the rotation of the Pacific plate around a different pole, suggesting a remarkably simple sequence of events: about 40 million years ago the motion of the Pacific plate shifted to a new pole of rotation and thereby changed its direction of migration, causing an abrupt kink in the Hawaiian chain.

Another tantalizing inference from geography suggests a possible confirmation of this theory. The Austral and the Tuamotu island chains also seem to bend sharply at an age of about 40 million years, and each is continued in a line of seamounts. These seamounts are parallel to the Emperor system, and they could have been formed by the ro-

tation of the plate around the same pole. For this conclusion to be accepted, however, it must be shown that the seamounts become progressively older to the north. As yet there are few dates established for the seamount series; those that have been obtained suggest a more complex interpretation.

The reconstruction of plate motions from the tracks of hot-spot volcanoes depends ultimately on the belief that the hot spots themselves are immobile or nearly so. This assumption appears to be justified. Minster and his colleagues have made accurate maps of relative plate motions by methods that do not rely on the hot-spot positions. Their work shows that prominent hot spots throughout the world have not moved in relation to one another during the past 10 million years. Other investigators have compared the positions of hot spots over much longer periods. Those determinations seem to show that groups of hot spots in one ocean have moved with respect to groups in other oceans over the past 120 million years—since the supercontinent Pangaea broke apart. This wandering of groups of hot spots, however, is slow compared with the shifting of the lithospheric plates.

The Population of Hot Spots

A census of the world's hot spots suggests that at least 122 have been active in the past 10 million years. Most of them meet all the particulars of the definition and can be classified without ambiguity. They are centers of volcanism that are not associated with plate boundaries and that form elevated domes with a diameter of up to about 200 kilometers. Also counted in the census, however, are several regions that lie on mid-ocean ridges or close to them; prominent among these are Iceland, the Azores and Tristan da Cunha, a small group of islands in the South Atlantic. The reason for the inclusion of these areas is that they seem more characteristic of hot spots than of normal mid-ocean ridges. The volume of material they have ejected greatly exceeds the norm for mid-ocean ridges; that is why they have been built up into islands while the rest of the ridge crest has remained submerged. More important, the lavas of these regions are the alkali-rich basalts that are rare at plate margins but that are typical of hot spots.

Our census probably underestimates the number of hot spots. There are domes or rises on some plates that are not capped by volcanoes; in spite of similarities in shape and geophysical properties we have not included them. There are probably also small, active volcanoes on the ocean floor that remain to be discovered. Finally, we have not attempted to include hot spots on con-

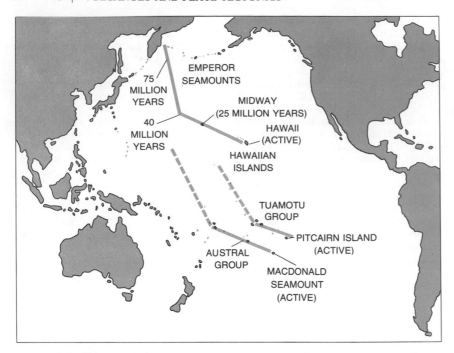

PACIFIC ISLAND CHAINS can be interpreted as tracks formed by the movement of the sea floor over stationary hot spots. The Hawaiian Islands grow older toward the northwest, beginning with Hawaii itself. Two other chains parallel to the Hawaiian Islands display a similar pattern of ages. They are the Austral group, which begins with the MacDonald Seamount, and the Tuamotu group, which begins with Pitcairn Island. All three chains could be generated by the same clockwise rotation of the Pacific plate. The age sequence of the Hawaiian Islands is continued in the Emperor Seamounts, which strike northward from a bend formed 40 million years ago. The change in direction implies that until then the rotation of the plate was centered on a different pole. Seamounts also extend to the north from the Austral and Tuamotu groups (*broken lines*), but there is no convincing evidence their ages form a linear sequence.

verging plate margins. In these areas volcanic activity is both abundant and complex, and it would be difficult to isolate the contribution of hot spots from other sources of volcanism. It should be noted, however, that basalts rich in alkali metals are found in some converging-plate zones.

Of the 122 hot spots we. have identified, 53 are in ocean basins and 69 are on continents. Among the oceanic hot spots there is a tendency to congregate on mid-ocean ridges: 15 lie on the crests of ridges and nine others are near the crests. The greatest concentration, however, is in Africa. The African plate has 25 hot spots on land, eight at sea and 10 more on or near the surrounding ocean ridges, for a total of 43.

Even allowing for possible errors in our census, the inhomogeneity is striking. The African plate constitutes 12 percent of the world's surface area, but it has 35 percent of the hot spots. The large-scale topography of the African continent is also unusual. It is characterized by basins and swells, and in recent epochs South and East Africa have been greatly uplifted to produce highlands and the Great Escarpment. The topography and the abundance of hot spots are almost certainly related. Both can be explained by the hypothesis that Africa

has come to rest over a population of hot spots.

The most compelling evidence that Africa is stationary is that at some hot spots lavas of several ages are superposed. If the continent were moving, of course, these lavas would be spread out in a chronological sequence. A few hot spots in the vicinity of Cameroon seem at first to be aligned like the island chains of the Pacific. It has been found, however, that these volcanoes are not arranged in chronological sequence. Their alignment is presumably a coincidence; it cannot have been caused by the motion of the plate.

Africa's basin-and-swell topography and the uplifting of large regions could be a direct result of the continent's immobility. Seismic studies have shown that the mantle is not homogeneous, and if there are variations in composition, there may be local concentrations of radioactive elements. The decay of these elements, which contributes a major part of the heat generated in the interior of the earth, would heat and expand some parts of the mantle more than others. The effects of the expansion might uplift regions of a stationary continent, but on a moving continent the uplift would be smeared out and would not be detected.

It is tempting to generalize from these observations, and it does seem there is a relation between the number of hot spots on a continent and the speed with which the continent is moving over the mantle. Antarctica, China and Southeast Asia, like Africa, have relatively large numbers of hot spots on land. Rates of sea-floor spreading imply that if Africa is stationary, these other regions are moving only slowly. In contrast, on rapidly moving plates, such as the North and South American ones, hot-spot volcanism is uncommon.

Opening of the Atlantic

Dated sequences of rock imply that Africa had numerous active volcanoes until the breakup of Gondwanaland 120 million years ago. The volcanic activity then ceased, and it did not resume until 30 million years ago. The two periods of activity and the long intermission between them can be read as signposts indicating the stages in the formation of the Atlantic Ocean.

The earlier episode of volcanic activity suggests that when Africa was a component of Gondwanaland, it was stationary over the mantle. When the supercontinent fractured along the present line of the Mid-Atlantic Ridge, Africa moved east. The motion over the mantle extinguished the volcanism for the next 90 million years. It is convenient to assume that the developing mid-ocean ridge was then stationary and that the two continents spread symmetrically away from it. They rotated in opposite directions around a pole near Cape Farewell on the coast of Greenland.

About 30 million years ago the African plate came to rest; volcanic activity on the continent resumed, and it has continued to the present. Although the African plate had stopped, sea-floor spreading had not. As a result the Mid-Atlantic Ridge was forced to begin drifting west. The relative motion of Africa and South America was unchanged, but the speed of the South American plate with respect to the mantle was doubled. When the Mid-Atlantic Ridge began its migration, the hot spots on the crest were left behind. Today a row of hot spots, which includes Tristan da Cunha and Ascension Island, is found a few

ATLANTIC HOT SPOTS also record the passage of the lithospheric plates. Several of these hot spots are on or near the Mid-Atlantic Ridge; a notable example is Iceland, which has been built up from the massive eruptions of volcanoes on the ridge crest. From some hot spots transverse ridges of volcanic rock extend back to the continental margins, indicating that the mid-ocean ridge developed over these volcanoes and that they were already active when the continents separated.

ARCTIC CIRCLE

JAN MAYEN ISLAND

ICELAND

FAROE ISLANDS

? ?

• 45° NORTH

AZORES •

COLORADO SEAMOUNT

MADEIRA •

MID-ATLANTIC
RIDGE

CANARY
ISLANDS

CAPE VERDE
ISLANDS

NORTH BRAZILIAN RIDGE

ST. PAUL'S
ROCKS

FERNANDO PO
PRINCIPE
SÃO TOMÉ
ANNOBÓN

EQUATOR

FERNANDO
DE NORONHA

ASCENSION
ISLAND

ST. HELENA •

• TRINIDADE

RIO GRANDE RIDGE

WALVIS RIDGE

TRISTAN DA CUNHA •

NIGHTINGALE

GOUGH ISLAND

DISCOVERY SEAMOUNT

BOUVET ISLAND

ANTARCTIC CIRCLE

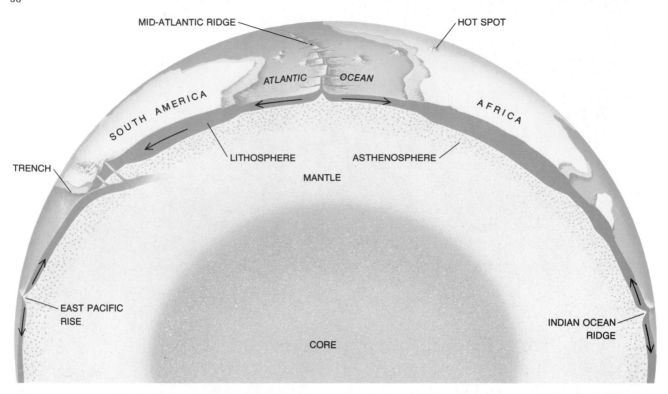

MID-ATLANTIC RIDGE

HOT SPOT

ATLANTIC OCEAN

SOUTH AMERICA

AFRICA

TRENCH

LITHOSPHERE

ASTHENOSPHERE

MANTLE

EAST PACIFIC RISE

INDIAN OCEAN RIDGE

CORE

MOVEMENT IN THE EARTH is described by the theory of plate tectonics. The lithosphere, the cool and rigid layer that includes the crust of the earth, is broken into about a dozen large plates. These plates move over the asthenosphere, a layer that is hotter and capable of slow deformation. The crust is the top of the lithosphere; the rest of the lithosphere and all of the asthenosphere are parts of the mantle. Lithospheric plates move apart as new material is added to them along mid-ocean ridges. Where two plates come together one dives under the other and is reabsorbed; this process, called subduction, results in extensive volcanic activity. Hot spots are small volcanic regions typical of neither mid-ocean ridges nor subduction zones. Unlike most other volcanoes, they are often found far from plate margins, and even when they are near a plate margin, they can be distinguished by the volume of lava they eject and by its composition.

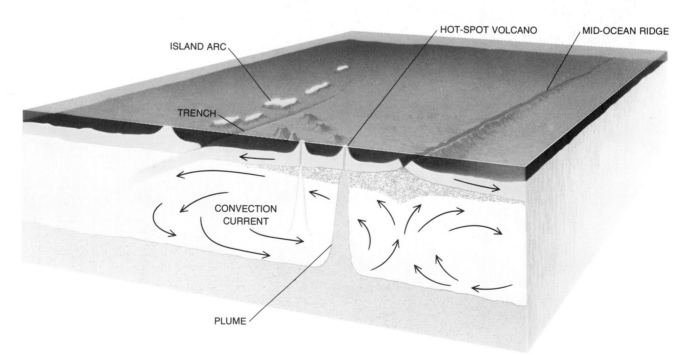

ISLAND ARC

HOT-SPOT VOLCANO

MID-OCEAN RIDGE

TRENCH

CONVECTION CURRENT

PLUME

SOURCE OF A HOT-SPOT VOLCANO is thought to be a "plume" rising from deep within the mantle. Differences in composition between the lavas ejected at hot spots and those characteristic of plate-margin volcanism suggest that the two kinds of lava come from different parts of the mantle; indeed, the source of the hot-spot lavas may have been isolated for as long as two billion years. Much of the mantle is probably stirred by convection currents, so that the plumes must originate in some region isolated from this circulation. For example, they might come from a stagnant zone in the middle of a convection cell, or from a layer below the reach of the mantle currents. A lithospheric plate moving over a plume leaves a trail of volcanoes that grow older with distance from present site of volcanic activity.

hundred kilometers east of the crest on lithosphere 30 million years old.

The evidence for the mobility of the Mid-Atlantic Ridge lies on the sea floor. From Tristan da Cunha a range of volcanic debris called the Walvis Ridge extends to the northeast. It is believed to be the track of the hot spot during the earlier part of the expansion (when the crest was fixed and Africa was moving), since it extends to lavas on the African coast that date from the disintegration of Gondwanaland. On the other side of the Mid-Atlantic Ridge another line of volcanic debris, the Rio Grande Ridge, extends to the Brazilian coast. There is no hot spot at its seaward terminus, which is separated from the mid-ocean ridge by a gap equivalent to 30 million years.

The disposition of these surface features can be explained by assuming that when the Atlantic was born, Tristan da Cunha was already an active volcano lying directly on the rift that opened to form the ocean. Lava from the hot spot overflowed onto both sides of the ridge and was rafted away by the spreading plates; continued eruption formed a V-shaped pair of tracks. When the mid-ocean ridge began to move west, the hot

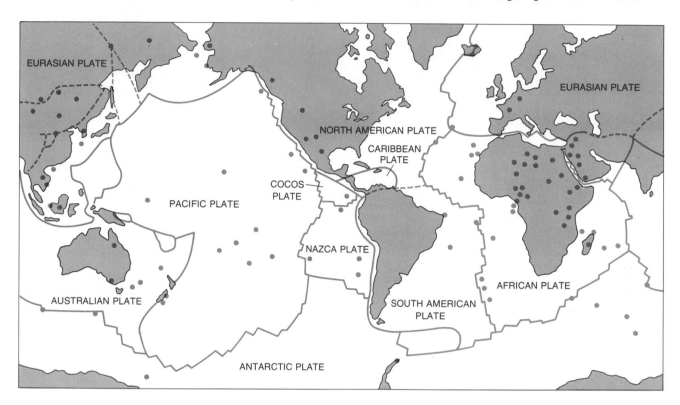

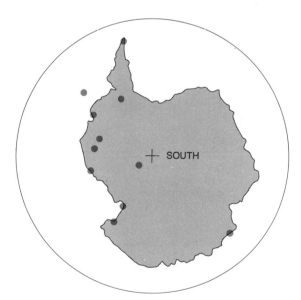

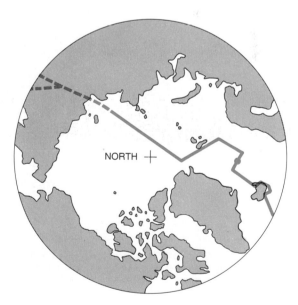

POPULATION OF HOT SPOTS includes at least 122 that have been active in the past 10 million years. They are found on all the major plates and on both oceanic and continental crust, but their distribution is decidedly nonuniform. There is a concentration along mid-ocean ridges, and in particular along the Mid-Atlantic Ridge; what is even more conspicuous, of the 122 hot spots 43 are on the African plate. Together with other evidence, this abundance of hot spots suggests that the African plate is stationary over the mantle. If the African plate is adopted as a frame of reference, other areas that have many hot-spot volcanoes, such as Antarctica and Southeast Asia, are found to be moving only slowly; on fast-moving plates hot-spot volcanism is rare. The map is based on one prepared by W. S. F. Kidd.

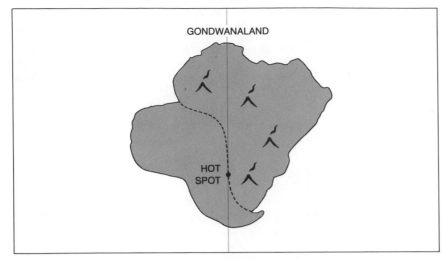

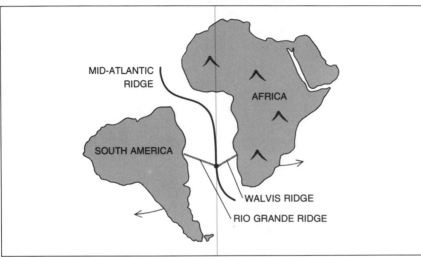

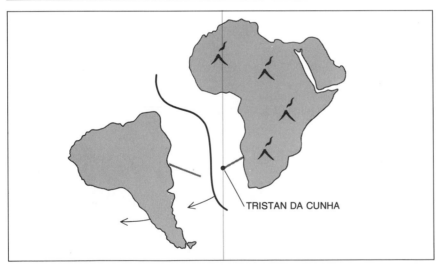

OPENING OF THE SOUTH ATLANTIC began 120 million years ago, when the great southern continent Gondwanaland broke apart. Until then hot-spot volcanoes had been abundant in Africa, suggesting that the continent was stalled over the mantle. When a fissure separated the continents, they receded symmetrically from the developing mid-ocean ridge and the motion of the African plate extinguished the hot spots. About 30 million years ago Africa came to rest and the present era of volcanism on the plate began. Because sea-floor spreading continued at the Mid-Atlantic Ridge, the ridge itself was forced to move west and the speed of South America was doubled. The ridge formed along a line that included several hot spots (*only one is shown*). As long as the ridge was stationary these hot spots generated trails of volcanic rock that extended back to the continental shores. When the mid-ocean ridge began its migration, the hot spots "fell off" the ridge crest and are now isolated on the African plate.

spot was left behind on the stationary African plate. It could no longer produce a lateral ridge; instead its successive lava flows simply piled up, one on top of the other. Today at Tristan da Cunha young volcanic rocks are found along with lavas at least 18 million years old. Since lava was no longer deposited on the American plate, the Rio Grande Ridge was also terminated.

Because hot spots are particularly common along ridges and seem to exercise some control over their location, it is reasonable to suppose that the crest of the Mid-Atlantic Ridge will someday jump back to the hot spots it has abandoned. If it does, the 30-million-year gaps in the Walvis Ridge and the Rio Grande Ridge will remain as a record of the interlude.

In the North Atlantic a somewhat different history can be read from the sea floor. The North Atlantic formed by the rotation of the Eurasian and North American plates around a pole in the Arctic Ocean. As we noted above, however, the American plate was already rotating around a pole in Greenland, near Cape Farewell, as a result of its separation from Africa. A single plate cannot rotate around two poles that are both fixed, and in this case the Arctic pole was itself in motion. The result was a shift in the position of the northern Mid-Atlantic Ridge.

When the North Atlantic began to open 80 million years ago, the locus of sea-floor spreading was west of Greenland. Spreading continued there until 50 million years ago and created Baffin Bay. An extinct hot spot left a pair of lateral ridges that trace this movement, extending to Disko Island in Greenland and to Cape Dyer on Baffin Island. Meanwhile 60 million years ago a new mid-ocean ridge developed east of Greenland. The continents have continued to diverge along that line since then.

Motion at Subduction Zones

We have seen that hot spots provide a method for translating the relative motion of lithospheric plates into motion with respect to the mantle. This frame of reference has been employed in clarifying an important aspect of the behavior of the plates that had been imperfectly understood.

When an oceanic plate collides with a continental one, the oceanic plate usually dives toward the mantle and is subducted. That is because the continental plates are thicker and more buoyant. The partial melting of the sinking plate leads to volcanic activity above the subduction zone, but this activity can have two quite different surface expressions. In some cases an island arc forms offshore. The most prominent examples of this process are in South Asia, where

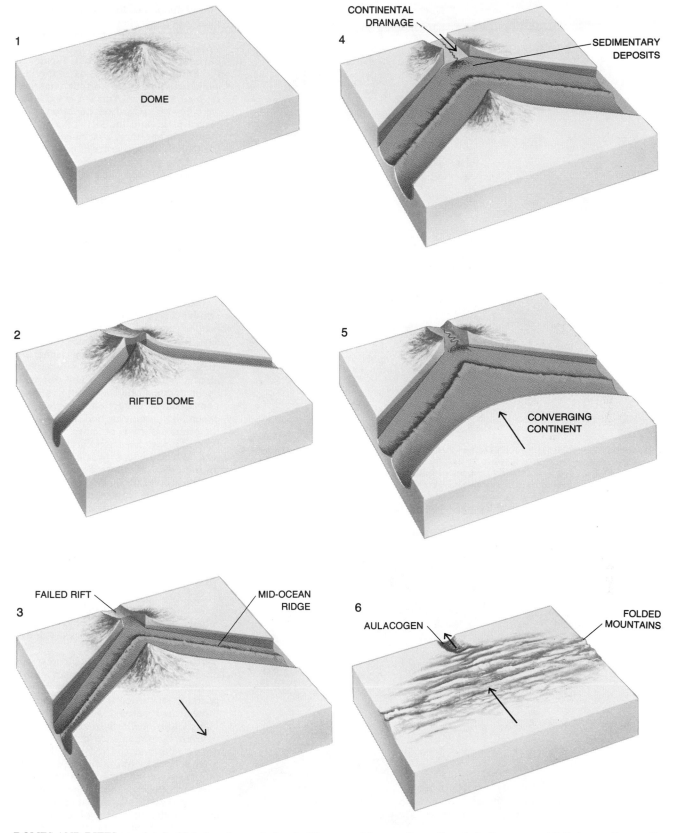

1 DOME

2 RIFTED DOME

3 FAILED RIFT — MID-OCEAN RIDGE

4 CONTINENTAL DRAINAGE — SEDIMENTARY DEPOSITS

5 CONVERGING CONTINENT

6 AULACOGEN — FOLDED MOUNTAINS

DOMES AND RIFTS associated with hot spots may be involved in the fracturing of continents and the opening of oceans. A dome, often capped with volcanoes (*1*), forms on a continent that is at rest over the mantle. Rifts develop in the dome (*2*), frequently in a three-armed pattern. Two of the arms widen and eventually become the basin of an ocean (*3*), but the third arm fails to develop further. This failed rift may become a major river valley draining the continent and transporting sediments to the new sea (*4*). Later another continent approaches the site of the original rift and closes the ocean (*5*). The collision pushes up a belt of folded mountains, reversing the drainage pattern and carrying sediments back into the failed arm of the rift. Eventually the rift is filled; what remains is a trough of deep sediments roughly perpendicular to the mountain belt (*6*). Nicholas Shatsky of the U.S.S.R. has named such features aulacogens.

RIDGES IN THE NORTH ATLANTIC and Arctic oceans suggest that the locus of sea-floor spreading in this region shifted between 50 and 60 million years ago. Initially the continents separated along a ridge west of Greenland, opening Baffin Bay. An extinct hot spot has left a record of this movement in trails of volcanic rock that extend from the dormant ridge to Cape Dyer on Baffin Island and to Disko Island in Greenland. About 60 million years ago sea-floor spreading began at the present site of the Mid-Atlantic Ridge, which passes east of Greenland.

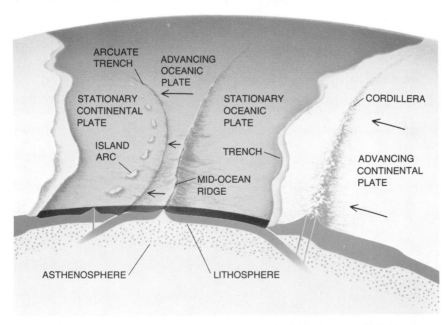

CONVERGENT PLATE MARGINS can assume two different forms. Where an oceanic plate is advancing on a stationary continent, the thin and flexible sea floor buckles offshore in a characteristic arcuate pattern; the volcanoes rising above the subduction zone create an island arc like the arcs of Japan and Indonesia. When a moving continent overrides a stationary oceanic plate, the descending slab of lithosphere is forced to bend at the coastline; as a result the volcanoes rise through the continent, forming a mountain system such as the Andes.

the subduction of the Indian-Australian plate has generated the Indonesian archipelago, and in East Asia, where the sinking Pacific and Philippine plates have produced the islands of Japan and the Philippines. In other cases the volcanic activity appears on the continental landmass. The Andes, for example, were thrown up by the subduction of the Nazca plate, and the Sierra Nevada of California and the Coast range of British Columbia derive from the subduction of the Pacific plate.

It has not been clear why the same process should have two dissimilar manifestations. By referring the plate motions to the hot spots we have attempted to resolve the question. Island arcs form when the continent is stationary over the mantle and the ocean floor moves under it; coastal mountain ranges are raised up when the continent overrides a stationary oceanic plate.

The only plausible explanation for the regular shape of island arcs has been proposed by F. C. Frank of the University of Bristol. He has pointed out that a flexible but inextensible thin spherical shell can bend in on itself only along a circular bend or fracture. This can readily be demonstrated by denting a pingpong ball. It is suggested that where the ocean floor is moving and free to adopt the preferred shape offshore islands are formed in the characteristic arcuate pattern. When the continent is advancing, on the other hand, the oceanic plate is submerged before it can develop an offshore arc. The known motions of the plates in the Pacific region support this conjecture. The oceanic plates of the Pacific are advancing on Eurasia and underthrusting it, but they are being overtaken by the Americas.

Doming and Rifting

So far we have considered hot spots mainly as indicators of plate motion. They may also act to initiate cycles of tectonic activity.

When a continent comes to rest, the dome that swells up over a hot spot is subject to fracturing. When a rift appears, it very often has a characteristic three-armed pattern. Forty years ago Hans Cloos, a German geologist, recognized the prevalence of such three-armed rifts and showed that they are often related to doming of the continental crust. We would suggest that these rifts are often the seed from which an ocean grows. It follows that the ultimate cause of the rupturing of a continent may be the continent's coming to rest over the mantle. The hot spots appear to guide the fracturing, although they are not necessarily its only cause.

The observed concentration of hot spots on mid-ocean ridges would be accounted for if this mechanism is commonly involved in the formation of oceans. The breakup of Gondwanaland accords well with this interpretation; it will be remembered that Africa was stationary until the disintegration began.

Typically two arms of the rift open to form an ocean basin, but the third arm fails and remains as a fissure in the continental landmass. By restoring the margins of the Atlantic Ocean to their positions before Pangaea split apart, an abundance of three-armed rifts is revealed. The successful arms merged to create the ocean, whereas the unsuccessful ones remained as rifts extending into the continents. The best example of such a failed rift on the Atlantic coasts is the Benue Rift, which strikes away from the Gulf of Guinea into equatorial Africa.

A much more recent and more conspicuous example can be observed today where the Arabian Peninsula is splitting away from Africa. The Red Sea and the Gulf of Aden both represent arms of a three-armed rift. The third, dry arm strikes into Ethiopia from the Afar Triangle. The symmetry of the pattern is remarkable. The fact that Africa has been stationary over the mantle for 30 million years and that it bears extensive evidence of doming and rifting suggests that we could be witnessing the early stages in the disintegration of the African continent.

Aulacogens

The present cycle of tectonic activity, which dates from the breakup of Pangaea, is not the only one in the earth's history. The recognition that hot spots, domes and rifts form a sequence in the fragmentation of continental landmasses has led to the discovery of a clue that could prove valuable in attempts to reconstruct the earlier wanderings of the lithospheric plates.

In 1941, as German forces threatened the main oil-producing area of the U.S.S.R., Nicholas Shatsky, a Russian stratigrapher, began a search for sedimentary basins that might contain new oil reserves. From the stratigraphic sequences compiled by Shatsky and his colleagues a previously unrecognized pattern emerged. Over much of the Russian and Siberian platforms the sedimentary layer is about a kilometer thick, but they found several narrow troughs up to 800 kilometers long where the sequence is three times the normal thickness. They named these formations aulacogens, from Greek words meaning "born of furrows." Aulacogens are rifts that extend from belts of folded mountains into continental platforms.

Aulacogens can now be recognized as failed arms of three-armed rift systems. When the two successful arms opened to form an ocean, the failed arm remained as a rift valley running inland from the

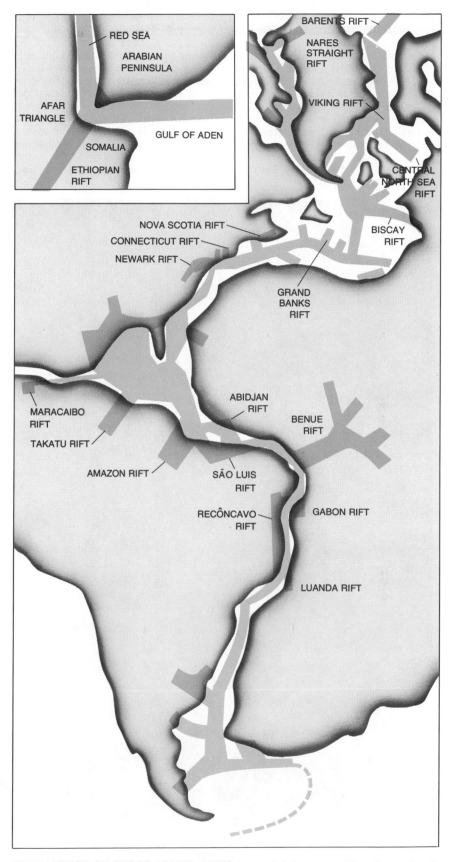

PREVALENCE OF THREE-ARMED RIFTS is revealed by reassembling the continents surrounding the Atlantic. In most cases two of the arms were incorporated into the Atlantic while the third remained a blind rift. A similar process can be observed today where the Arabian Peninsula is splitting away from Africa (*upper left*). The Gulf of Aden and the Red Sea form two branches of a rift; the third extends from the Afar Triangle into Ethiopia and Somalia.

new seacoast. The rift became a feature of the drainage pattern of the continent, accumulating a thick deposit of sediments. Later another continent approached the coast, closing the ocean and blocking the rift. Compressional forces generated by the collision pushed up a chain of folded mountains. The remnant of the rift was a deep bed of sediments striking almost perpendicularly to the mountain chain.

The aulacogens that Shatsky recognized in the U.S.S.R. were of Paleozoic age (between 225 and 600 million years old). Paul Hoffman of the Geological Survey of Canada has since described a formation called the Athapuscow aulacogen that is two billion years old; it underlies the eastern arm of Great Slave Lake in northern Canada. Shatsky himself recognized what is probably the best-developed aulacogen in North America. It is a bed of sediments 15 kilometers deep in southern Oklahoma, parallel to the Texas border. It formed as a rift 600 million years ago when an ocean opened up roughly where the North Atlantic is today. The closing of that ocean was responsible for the building of the Caledonian, Appalachian and Ouachita mountains.

These ancient aulacogens are evidence that the cycle of continental disintegration and reassembly has been going on for at least two billion years. The development of domes and rifts in continents that come to rest over hot spots may have been a part of the process throughout this period.

AULACOGEN in southern Oklahoma is a remnant of an earlier cycle of continental drift. The photograph is a false-color image made in December, 1972, by the LANDSAT **satellite. The aulacogen starts in the belt of flatland in the lower half of the photograph and extends another 400 kilometers to the west. To the north are the Ouachita Mountains. When a sea opened up to the south and east 600 million years ago, the aulacogen was the failed arm of a three-armed system of rifts. The closing of that sea raised up the Ouachita range as well as the Appalachians. Erosion of the Ouachitas has added to sediments already in the rift to form a layer of sediments 15 kilometers deep.**

II

VOLCANIC PRODUCTS: LAVA, ASH, AND BOMBS

VOLCANIC PRODUCTS: LAVA, ASH, AND BOMBS

II

INTRODUCTION

Ejecta from volcanoes have a remarkably wide range of chemical, mineralogical, and physical properties. However, except for the unique sodium carbonate lavas from Ol Doinyo Lengai Volcano in East Africa, nearly all volcanic products are silicate rocks composed mainly of oxygen, silicon, and aluminum, with lesser amounts of iron, calcium, magnesium, sodium, titanium, and potassium. Basalts are relatively low in silica and relatively high in calcium and magnesium. Magmas that are more siliceous—andesites, dacites, and rhyolites—have more sodium and potassium and less magnesium and iron.

Magmas of different composition can form by differing degrees of partial melting of mantle rocks; by partial crystallization, which enriches the remaining melt with silica; and by melting and assimilation of crustal rocks of various composition.

Partial melting begins with the constituents in the source rock which have the lowest melting temperatures, leaving the more refractory components as solids. A small percentage of partial melt will thus tend to have a higher ratio of the easily melted constituents such as potassium than similar source rock that has been half-melted.

Partial crystallization works in the opposite direction. Minerals with high melting temperatures, such as olivine, precipitate early in a slowly cooling batch of magma and thereby change the composition of the remaining melt. When erupting magma rises to the surface it incorporates various rock types along its path of ascent. This contamination of the magma by its host rock changes the composition of the melt in a way that depends on the types and the amounts of rocks assimilated.

Physical products of volcanic eruptions include gases, liquids, and solids. Explosive volcanoes generate large volumes of gases and hot, but solid, rock fragments. Rapid gas expansion solidifies the rock fragments. The resulting mixture of hot, solid particles and gas can either billow upward as a huge ash cloud or avalanche rapidly down slopes as a fluidized flow. In less violent eruptions, lava may boil out in spectacular fire fountains or simply flow from cracks or vents into streams of lava. The main factors controlling the physical nature of volcanic products are the viscosity of the magma, its gas content, the rate of emission, and the environment of the vent. Submarine, subglacial, and subaerial eruptions of the same magma can produce vastly different rock deposits.

Fragmental volcanic products are called *pyroclastic*, which literally means fire fragments. They are erupted and deposited in two different ways; either as airfall deposits (*tephra*) or as pyroclastic flows. Airfall deposits form distinct layers that conform to the slopes of the ground surface on which they

Volcanic Products

Form	Name	Characteristics (dimensions)
Gas	Fume	
Liquid	*Lavas*	
	Aa	rough, blocky surface
	Pahoehoe	smooth to ropy surface
Solid	*Airfall fragments*	
	dust	$<\frac{1}{16}$ mm
	ash	$\frac{1}{16}-2$ mm
	cinders	$2-64$ mm
	blocks	>64 mm solid
	bombs	>64 mm plastic
	Pyroclastic flows	hot fluidized flows
	Mudflows	flows fluidized by rainfall, melting ice and snow, or ejected crater lakes

(From Decker and Decker, *Volcanoes,* p. 108, W. H. Freeman and Company, copyright © 1981.)

fell. The falling debris is generally sorted with coarse and heavy particles depositing first and nearby, while the dust is winnowed away to fall last, sometimes at great distances from the vent. Airfall deposits can be recognized by their characteristic layering and sorting.

Pyroclastic flows form when the fluidized emulsion of volcanic fragments and hot gases is too heavy to rise into an ash cloud. Instead it forms a glowing avalanche, also known as a *nuée ardente*. A hot fluidized avalanche can travel at high speeds, sometimes exceeding 100 kilometers per hour. Small glowing avalanches often flow down valleys on a volcano's flanks, but a larger mass expelled at high velocity or accelerated by a steep slope can sweep over hills or across large flat areas for many tens of kilometers. After the avalanche loses its velocity, pyroclastic flow deposits tend to pond in low-lying areas. These deposits are distinct from airfall debris; they exhibit only vague layering with almost no sorting of the fine and coarse fragments, and have the massive appearance of concrete.

Many subduction-zone volcanoes have eruptions that begin with pyroclastic emissions and end with thick, viscous lava flows. This sequence may be caused by higher concentrations of gas in the upper parts of their magma chambers before the eruption.

Oceanic rift volcanoes and Hawaiian volcanoes erupt mainly lava flows, with individual flows forming long narrow tongues. On the Island of Hawaii a typical lava flow might be 10 kilometers long, 200 meters wide, and 3 meters thick. Sometimes the central feeding channel of a lava flow crusts over and forms a tunnel filled with a rapidly moving stream of lava. As an eruption wanes, the lava in the tunnel drains downslope, leaving an empty cave known as a lava tube. These tubes are generally 1 to 10 meters in diameter and meander underground for hundreds of meters.

Volcanic gases—steam, carbon dioxide, sulfur dioxide, hydrochloric acid, and traces of others—are dissolved in magma. They exsolve as magma rises to the surface and pressure decreases. The texture of lava and volcanic ash is largely controlled by the number and size of gas bubble holes formed in the erupting material. Pumice is at one extreme; consisting mainly of holes, it will float on water. Dense solid rock without holes is the other extreme. Most volcanic rocks are somewhere in between.

The volumes of lava and pyroclastic rocks produced in individual historic eruptions range from a few cubic meters up to 20 cubic kilometers. On the average, if added together the volcanoes of the world's subduction zones produce about 1 cubic kilometer of new volcanic material each year, mostly pyroclastic. Rift volcanoes generate about 2.5 cubic kilometers per year, mainly submarine flows of pillow basalt. Hot-spot volcanoes produce about 0.5 cubic kilometers per year, largely basalt lava flows in oceanic settings, and lava flows or siliceous pyroclastic material in continental settings. The Earth's volcanoes thus have an average total output of about 4 cubic kilometers of new rock each year. However, many years go by without large volcanic eruptions until a Krakatau or Katmai-type eruption comes along to catch up on the lag. But the giant volcanic eruptions of historic record are dwarfed by the volumes of prehistoric volcanic rocks that seem to have been poured out in single enormous eruptions. Volumes on the order of 100 to 1000 cubic kilometers are not uncommon in many volcanic regions of the Earth. Historic time is too short to give a representative sample of the enormous power of volcanic activity.

4

The Lava Lakes of Kilauea

by Dallas L. Peck, Thomas L. Wright,
and Robert W. Decker
October 1979

*The eruptions of the Hawaiian volcano leave pools of
molten basalt that can take as long as 25 years to
solidify. They provide a natural laboratory for studying
the nature of magma from the earth's mantle*

Magma—molten rock—from the interior of the earth is responsible for a host of phenomena at the earth's surface. The flow of magma out of the mid-ocean rifts adds to and pushes apart the rigid plates that make up the earth's surface and carry the continents on their backs. All igneous rocks are by definition formed by the congealing of magma. If the magma is erupted at the surface as lava, it forms extrusive igneous rocks such as basalt; if it slowly crystallizes below the surface, it forms intrusive igneous rocks such as granite. In spite of the importance of magma, however, there is much that is not known about it. Most studies of the cooling, crystallization and other properties of magma have centered on the laboratory analysis of small samples and on theoretical extrapolations from already solidified lava. A group of us at the U.S. Geological Survey, Sandia Laboratories and several universities have taken a different approach. We have studied molten and solidifying lava in situ by examining three "lakes" of lava left in the wake of eruptions of the volcano Kilauea on the island of Hawaii.

The three lakes are filled with basaltic lava. Basalt is the commonest rock formed by the solidification of magma extruded to the surface of the earth, the moon and perhaps other bodies in the solar system. Rich in calcium, magnesium and iron, basaltic lava crystallizes to form dark rocks consisting mostly of the silicon oxides plagioclase feldspar, pyroxene and olivine (whose gem form is peridot), the iron oxide magnetite and the iron–titanium oxide ilmenite. Basalt is found on all the continents and covers huge expanses of land such as the Columbia River and Snake River plateaus in the northwestern U.S. and the Deccan plateau in western India. Basaltic lavas, erupting from the mid-ocean rifts to create the floor that underlies the sediments of the ocean basins, have poured forth throughout geologic time from the early Precambrian to the present. Although basalts vary significantly in chemical and mineral composition, they have all formed at high temperatures. In studying the lava lakes we undertook, among other things, to examine the feasibility of obtaining geothermal energy directly from deep bodies of basaltic magma. In principle the high temperature of the molten rock makes it attractive as a source of energy, although in practice numerous obstacles stand in the way of tapping its heat.

Most of the more than 500 active volcanoes on the earth are entirely or predominantly basaltic, including the active volcanoes that make up the southern two-thirds of the island of Hawaii: Mauna Loa, Hualalai and Kilauea. Mauna Loa is the largest of the three, rising four kilometers above sea level and almost nine kilometers above the adjacent ocean floor. Kilauea is the smallest, rising only 1.3 kilometers above sea level. Shaped like an inverted saucer, Kilauea is indented at the summit by a caldera, or large crater, from which radiate two rift zones. The eruptions of Kilauea are usually limited to the caldera and the rift zones, particularly the eastern rift zone and Halemaumau, the "fire pit" in the caldera. Since 1952 Kilauea has erupted on the average at least once a year.

The basaltic magma that feeds the eruptions comes from the earth's mantle at depths of at least 50 kilometers below the surface. Geological and geophysical data suggest that magma rising from these depths is stored in an irregularly shaped reservoir about four kilometers below the summit of Kilauea. In the formation of a lava lake lava from the reservoir erupts to the surface and flows into a depression. Eventually the natural dikes that channel the lava into the lake collapse, and so the lake is cut off from a source of lava and starts to solidify.

The lava lakes on Kilauea that we investigated were one formed in 1959 in Kilauea Iki Crater, a second formed in 1963 in Alae Crater and a third formed in 1965 in Makaopuhi Crater. The depth of the lakes is substantial, so that it takes a long time for them to cool and solidify. The times range from 10½ months for the shallowest lake (in Alae Crater) to about 25 years for the deepest (in Kilauea Iki Crater). The lakes differ not only in depth but also in how they were filled, which affects their temperature, homogeneity and structure.

Kilauea Iki lava lake, which reaches depths of 120 meters, was formed in a spectacular event characterized by 17 separate eruptions over 36 days. Between eruptions volcanic activity stopped and some lava flowed out of the lake through a rift halfway up the side of the depression. Fragments of crust that had formed earlier then sank, making the upper levels of the lake markedly inhomogeneous. The eruption temperatures of between 1,200 and 1,215 degrees Celsius were high, and the abundant magnesium oxide favored the formation of olivine.

Alae lava lake, which is 15 meters deep, is poor in olivine. Its lava formed in the course of a three-day eruption at temperatures of between 1,140 and 1,160 degrees C. Makaopuhi lava lake is intermediate between the other two lakes in depth, temperature of formation and olivine content. Reaching a maximum depth of 83 meters, the lake formed in a 10-day eruption at temperatures of between 1,150 and 1,200 degrees. Our investigation of Alae and Makaopuhi lakes was abruptly halted in 1969 by new eruptions that inundated the lakes with fresh lava. Kilauea Iki lake has remained as it was, and it is still being investigated.

As the lakes cooled and solidified after they had been cut off from their source of lava we examined the process by drilling more than 50 holes in them between 1960 and 1977 and taking core samples of both solid and molten material. At first we worked with a small portable rotary drill powered by a chainsaw motor. Later we switched to a large trailer-mounted drill rig. The drilling re-

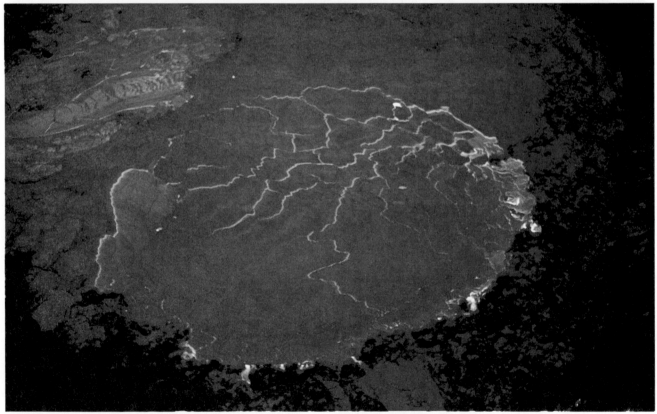

FOUNDERING OF SOLIDIFIED LAVA at the surface of Makaopuhi lava lake on Kilauea is shown in these photographs made in 1965. Gases exsolved from the underlying molten lava rose toward the surface and were trapped under the solidified but hot crust of the lake. In the top photograph the trapped gas gave rise to a gravitationally unstable situation that resulted in the foundering of a small piece of crust near the center of the lake's surface. The sunken lava was replaced by glowing molten lava (*center*). In the bottom photograph, made 10 minutes later, the foundering has extended to all the margins of the active part of the lake except the area at the upper right. The lava above 1,200 degrees Celsius is yellow. Between 900 and 1,100 degrees the lava is red, and below 900 degrees it is dark red or black.

vealed an interesting structure in depth. Like a freezing lake of water, a lava lake solidifies from the top down. The surface of a lava lake is cooled by air and particularly by rain, which falls copiously on Hawaii. (Over a four-year period 10 meters of rain fell on Alae lava lake.) Unlike a water lake, however, a lava lake also solidifies from the bottom up. That happens because the rock under a lava lake is cooler than the molten

lava. As a result the molten lava is sandwiched between two layers of solidified crust and takes the shape of a lens.

The crust at the bottom of the lake is always thinner than the crust at the top because the rock under it has a low thermal conductivity and is not rapidly cooled by rainwater. The molten lens decreases in thickness as the lake solidifies from the top down and from the bottom up. A lava lake, like a lake of

water, is shallower at the edges than it is at the center, so that the top and bottom crusts fuse at the edges of the lake, separating the molten lens from the rock enclosing the lake basin. As the top and bottom crusts continue to thicken the lens decreases in diameter and thickness until it disappears and the lake becomes a single body of solidified lava.

The ice on a lake of water is quite distinct from the water below it. That is

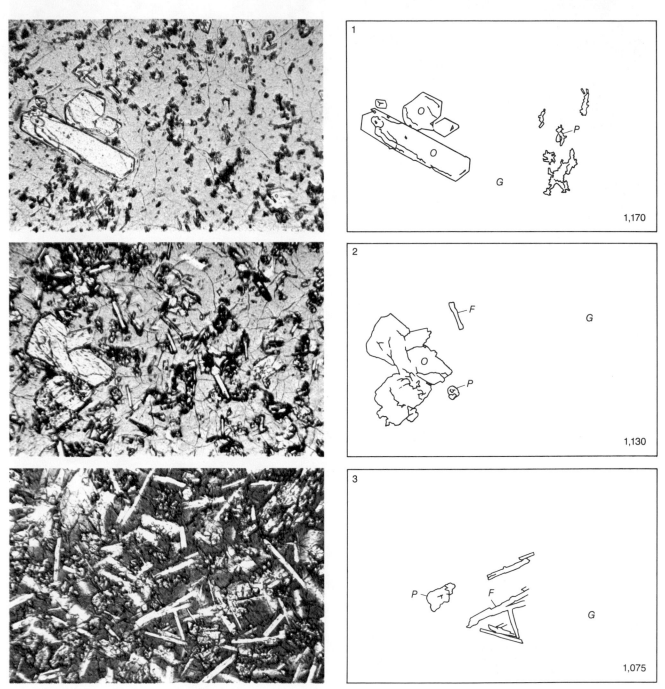

PHOTOMICROGRAPHS OF BASALTIC LAVA quenched from different temperatures in Makaopuhi lava lake show how the lava crystallizes. At the right of each photograph is a map identifying the minerals in the lava and a label indicating the temperature of the lava in degrees C. The minerals include the silicon oxides plagioclase feldspar (*F*), pyroxene (*P*) and olivine (*O*), the iron oxide magnetite (*M*), the iron–titanium oxide ilmenite (*I*) and the calcium fluoride–phos-

phate apatite (*A*). The maps also mark the location of lava in the glassy state (*G*): a supercooled liquid in which the silicon oxide molecules have not been organized into crystals. In photomicrographs *1* and *2*, which represent fluid lava at an intermediate depth in the lake, the crystals are sparse, dominated by coarse crystals of olivine (*O*). As the temperature of the lake decreases with decreasing depth, fine crystals of pyroxene (*P*) and pale, lathlike crystals of plagioclase feld-

not the case with a lake of basaltic lava, which consists not of one chemical compound but of different minerals that crystallize at different temperatures and rates. The solidified top crust grades slowly downward into the fluid lava through a region of partially molten crust several tens of centimeters thick. Within this region the temperature and the ratio of crystal grains to melt increase smoothly with depth. There is an interface, however, across which the physical properties of the partially molten lava change sharply. Above the interface the lava is solid enough to be drilled. Below the interface the lava is a fluid that yields like taffy when a drill probe is pushed into it. At the interface, whose temperature is 1,070 degrees C., crystal grains and melt are equally abundant. At 980 degrees the lava is entirely crystallized except for a small fraction in the glassy state: a super-cooled liquid in which the silicon oxide molecules have not been organized into crystals.

The rate of thickening of both the top and the bottom crust decreases with time because the solidified material is a poor conductor of heat and so acts as an insulator. The top crust of all three lakes reached a thickness of .3 meter after a day and two meters after a month. After

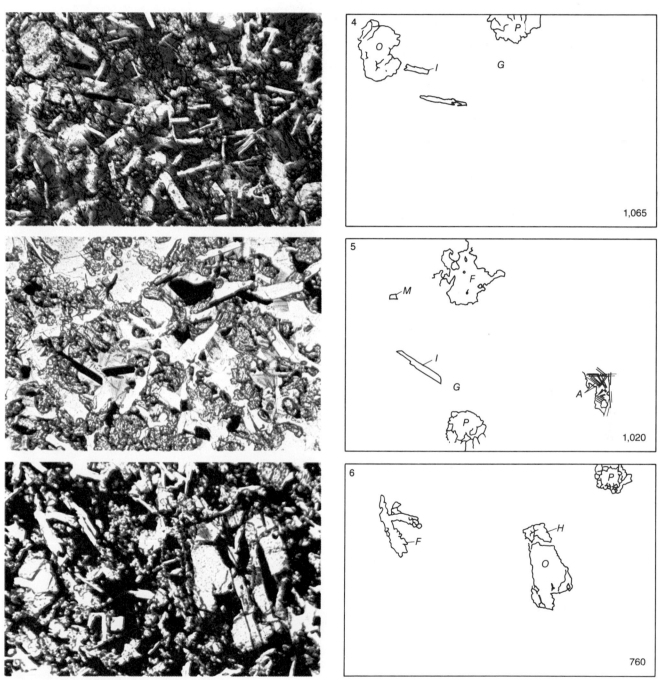

spar (*F*) form in increasing abundance. In photomicrographs *3* and *4*, made from the solidified lava near the base of the crust of lava at the top of the lake, the crystals of pyroxene (*P*) and plagioclase feldspar (*F*) are larger and commoner, constituting about 50 percent of the lava. In photomicrograph *4* the first crystals of ilmenite (*I*) can be seen as opaque blades. In photomicrograph *5* the lava is mostly crystal, consisting of an interlocking network of crystals of pyroxene (*P*), plagioclase feldspar (*F*) and ilmenite (*I*) surrounding small quantities of a pale brown glass (*G*) containing needles of apatite (*A*) and a few tiny cubes of magnetite (*M*). In photomicrograph *6* the lava is entirely solid but not completely crystalline; it still consists of a few small pools of silicon-rich glass. The lavas of Alae and Kilauea Iki lakes are similar in composition to the lava of Makaopuhi lake, although there are nonetheless significant chemical differences among them.

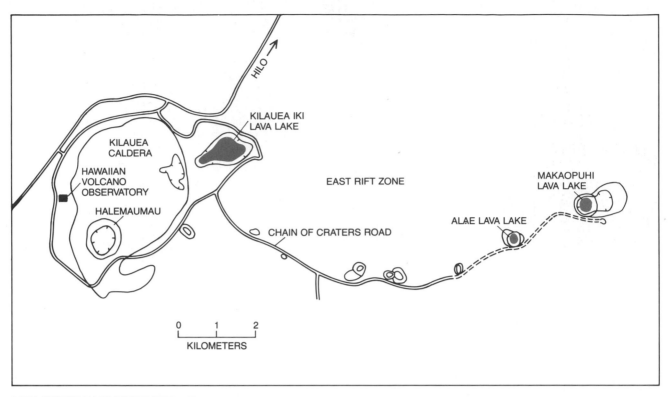

MAP OF THE EAST RIFT ZONE of Kilauea shows the location of the three lava lakes the authors investigated: one formed in 1959 in Kilauea Iki Crater, a second formed in 1963 in Alae Crater and a third formed in 1965 in Makaopuhi Crater. Halemaumau is the fire pit in the caldera at the summit of the volcano, which is shaped like an inverted saucer. Over the past 10 years fresh lava flows have covered Alae and Makaopuhi lava lakes and the part of the Chain of Craters road indicated by the broken lines. Kilauea Iki lake was not covered.

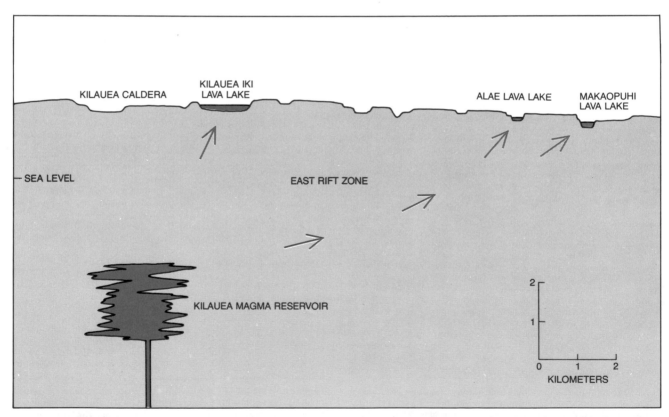

MAGMA RESERVOIR lying about four kilometers below the summit of Kilauea is the source of the lava that erupted to the surface and flowed into depressions to form the lava lakes. Kilauea Iki lava lake, which reaches depths of 120 meters, was formed in a spectacular event characterized by 17 separate eruptions over 36 days. Makaopuhi lava lake, which has a peak depth of 83 meters, was formed in an eruption that lasted for 18 days. Alae lava lake, which is 15 meters deep, was created in an eruption that lasted for three days.

a year the top crust of the two deepest lakes was eight meters thick. In 1976, 17 years after Kilauea Iki lava lake was formed, the top crust was 45 meters thick. From our core-sample data we found the rate of increase in thickness to be proportional to the square root of time until the maximum temperature in the lake falls appreciably below the initial temperature at the center of the lake. The thickness in meters is approximately equal to .4 multiplied by the square root of the time in days.

Since with the passage of time the top crust gets thicker at a lower rate, its temperature must also decrease at a lower rate. Measurements made with steel-sheathed thermocouples inserted into the drill holes indeed show that the rate of decrease in the temperature at any given depth in the top crust is lowered with time. With the aid of a computer we developed a numerical model for predicting the temperature at any depth in a lava lake. The model is based on Fourier's law of heat conduction: The flow of heat between two regions is proportional to the difference in temperature between those regions. The model incorporates the external processes that affect the flow of heat in a lava lake (cooling by rainwater), the latent heat of basalt (the heat released by crystallization) and variations in the density, porosity and thermal conductivity of the lava.

We applied this numerical model to the top crust of Alae lava lake, where we had regularly measured the temperature over a four-year period during which the lake had solidified and cooled to less than 100 degrees C. The average difference between the calculated temperature and the measured temperature was less than two degrees. Such close agreement tends to confirm the validity of the model, the main import of which is that the vaporization of rainwater from the hot crust of a lake has a large effect on the lake's decrease in temperature.

As molten lava cools, gas in the melt is driven out of solution. The gas either escapes into the atmosphere or remains in the lava as vesicles, or bubbles. The vesicles that were formed at the time of the eruption of the three Kilauea lava lakes and were frozen into the crust at shallow depths are chiefly spheres as much as a centimeter in diameter. They were apparently created at high temperatures when the lava became supersaturated with gas on the reduction of the confining pressure above it, just as bubbles appear in a bottle of soda water when the cap is removed. With increasing depth in the lava the confining pressure increases, and so the vesicles become smaller and scarcer. Below six meters in Alae lava lake most of the vesicles are minute angular pores less than a millimeter in diameter. They were ap-

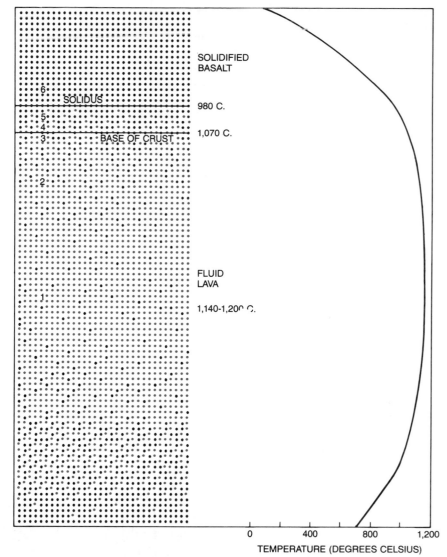

RATIO OF CRYSTAL TO MELT in a lava lake is shown in this highly schematic cross section (*left*). A lava lake solidifies from the top down and the bottom up. At temperatures below 980 degrees C. the lake is entirely solid. Between 980 and 1,070 degrees the lava is partially molten, but it is rigid because the melt is trapped in the interstices of a crystalline network of solidified lava. At 1,070 degrees the melt (*color dots*) and the crystal grains (*black dots*) are equally abundant. Above 1,070 degrees the lava is a fluid that yields like taffy when a drill probe is pushed into it. The numbers correspond to the positions of the lava shown in the photomicrographs on pages 50 and 51. At the right is a graph of the temperature as a function of depth.

parently created when gas was driven out of solution by the crystallization of the cooling lava. The composition of the gases expelled from the lava also changed as the lake cooled: water vapor increased in abundance at the expense of the more rapidly exsolved gases of carbon and sulfur compounds.

Although a solidifying lava might be expected to sink because its crystals are denser than the melt, the lava in Alae lake became more buoyant as it solidified because it was filled with gas vesicles. Therefore as the lens of molten lava below the surface at the center of the lake solidified it pushed up the surface above it. At the edges of the lake, where there was no solidifying lens because the top and bottom crusts

had fused, the surface subsided as the cooling lava thermally contracted and became denser; since this lava had already solidified, no bubbles were being formed in it that would make it more buoyant. The entire surface of Kilauea Iki and Makaopuhi lakes, on the other hand, has generally subsided as the lakes have cooled. Since these lakes are deeper than Alae lake, the higher pressures within them have hindered the formation of vesicular lava that would have pushed up the surface.

We investigated the crystallization of basalt by taking core samples of the semimolten lava at the base of the top crust, of the molten lava below the top crust and of the pumice and molten lava erupted to the surface when the lakes first formed. The samples enabled us to reconstruct the history of the crystallization of a lava lake. At first the crystals in the fluid lava are sparse, dominated by coarse grains of olivine. As the temperature of the lake decreases, fine crystals of pyroxene and pale, lathlike crystals of plagioclase form in increasing abundance. Near the base of the crust, at 1,070 degrees C., the crystals of pyroxene and plagioclase are larger and commoner, constituting about 50 percent of the lava. At 1,075 degrees iron and iron–titanium oxides are still in the glassy state, but at 1,065 degrees a few opaque blades of the iron–titanium oxide ilmenite can be seen. The abundance of such oxides gives the glass a rich brown color. Lava that has cooled to 1,020 degrees is mostly crystalline: an interlocking network of crystals of pyroxene, plagioclase and ilmenite surrounding small quantities of a pale brown glass containing needles of apatite and a few tiny cubes of magnetite. At 760 degrees the lava is entirely solid but not totally crystalline; it retains a few small pools of a silicon-rich glass.

The lavas of the three lakes are similar in composition, but there are nonetheless significant chemical differences among them and significant chemical variations within them. Three processes are responsible for these differences and variations. The first is the sinking of magnesium-olivine crystals, which are denser than the basaltic melt. The crystals migrate downward like grains of sand dropped into a pond. The effect of the migration is particularly apparent in an old solidified lake exposed in the walls of Makaopuhi Crater, where olivine is sparse at the top of the lake but abundant between six and 24 meters above the bottom. Most of the variation in the chemical composition of lava with more than 6 percent magnesium oxide is the result of variations in the abundance of olivine.

The second process, a separation of crystals of plagioclase and pyroxene in flowing lava resulting from differences in their density and shape, gives rise to subtler shifts in chemical composition. We first became aware of this separation in Makaopuhi lava lake when molten lava that had flowed into the drill casing was found to contain fewer crystals, particularly crystals of pyroxene, than the untapped fluid lava. The third process, which comes into play at lower temperatures, is the separation of interstitial melt from the network of crystals in the rigid but incompletely crystallized lava near the base of the top crust. Such melt fills fractures in the crust and oozes into drill holes.

The composition of the solidified but still hot basalt of the top crust of the lakes is also altered by the action of gases formed in the underlying crystallizing

lava that rise through fractures in the crust. The most striking alteration is the formation of red films of the iron oxide hematite in olivine and the replacement of silicate minerals by oxides at temperatures of between 500 and 800 degrees C. This alteration is due to variations in the concentration of free oxygen at different depths in the lakes. Where the oxygen is abundant the most oxidized form of iron, namely hematite, is highly stable. Where the oxygen is less abundant the less oxidized form of iron, namely magnetite, is stable. (The lack of free oxygen on the moon accounts for the absence there of hematite and the other principal oxide of iron, magnetite. On the moon the least oxidized form of iron, namely metallic iron, is stable.) At

temperatures of between 550 and 750 degrees the concentration of oxygen was high and the accumulation of hematite was quite conspicuous. The high concentration of oxygen presumably resulted in part from the decomposition of water vapor in the gases rising through the fractures.

Another phenomenon that alters the cooling crust of the lava lakes is that large cracks develop in the crust as the cooling basalt contracts. Such cracks open a minute or so after incandescent lava appears at the surface in the course of an eruption. As the crust continues to cool and thicken, the cracks propagate downward by further fracturing and new cracks open that divide the surface

DRILL RIG powered by a chain-saw motor sits on the surface of Makaopuhi lava lake. The rig, which was lowered onto the surface by a helicopter, gathered samples of solid and molten lava from various depths in the lake. In front of the drill rig is a water tank, which supplied water to quench the samples of hot lava. The crater wall behind the drill rig exposes an older columnar-jointed lava lake that was subsequently covered by several thin horizontal lava flows.

of the crust into polygons. Within a few hours the polygons become deformed as their centers are elevated by the vesicular expansion of the solidifying lava under them. Then still more cracks open, which are later characterized by sulfur and gypsum deposited in them by rising gases. Repetitive mapping of the cracks on the surface of Alae lava lake indicates that the rate of cracking is greatest at times when the crust is being chilled by a heavy rain.

The Kilauean lava lakes provide a good opportunity to study the physical properties of basaltic magma. Herbert R. Shaw of the U.S. Geological Survey and two of us (Peck and Wright) measured the viscosity of fluid lava both in the laboratory and in Makaopuhi lava lake. Viscosity is defined as the degree to which a fluid resists flowing under an applied force. We placed a stainless-steel viscometer, a rotor that measures viscosity by resistance to its turning, in the drill hole at depths of between 2.6 and 3.2 meters within the melt. The lava there was at a temperature of between 1,130 and 1,135 degrees C. and contained 25 percent crystal and between 2 and 5 percent gas bubbles. Our results confirmed those obtained in the laboratory, where the viscosity rose from 100 poises at 1,300 degrees to about 320 poises at 1,200 degrees. (A poise is a dyne-second per square centimeter.) This change is comparable to the difference in viscosity between heavy oil and stiff honey at room temperature. At temperatures below 1,200 degrees the viscosity increases sharply because of the accumulation of crystals in the melt.

The presence of gas bubbles at these temperatures made the lava plastic rather than fluid. Fluid behavior is characterized by deformation under the smallest applied stress, plastic behavior by deformation only after the application of a certain minimum stress. At low rates of rotation of the viscometer the lava at 1,130 degrees had a viscosity of 7,500 poises. The gas bubbles "stiffen" the lava just as air stiffens egg whites when they are beaten. At higher rates of rotation the bubbles elongated to form planes of weakness that lowered the viscosity to 6,500 poises.

The investigations of the lava lakes in the 1960's were undertaken to discover more about the nature of magma. Over the past four years work on the lava lakes has intensified as a part of the search for alternative sources of energy. The magma underlying the U.S. could conceivably provide 13×10^{21} calories of heat per year, an amount equivalent to what would be released by the burning of 9,000 billion barrels of oil.

The obstacles to tapping this energy are large. Methods have to be developed for finding a magma body and determining its size, shape, depth and heat content. Techniques are needed for drilling into a deeply buried magma body, putting a heat-transfer device into it and extracting the energy through the hole. The recovery of magma energy also presents a challenge to the materials sciences: materials must be found that can operate successfully for long periods in the hot, pressurized and corrosive environment of a magma.

Kilauea Iki lava lake provides a natural laboratory for work aimed at overcoming these obstacles. The lava in the lake was partially outgassed in the course of the eruption and is under relatively low pressure, so that the lake is not entirely representative of deeply buried magma. The magma at these shallow depths is much more accessible, however, and hence we have been able to apply what we had already learned about the physical and chemical properties of the melt in the lake to test poten-

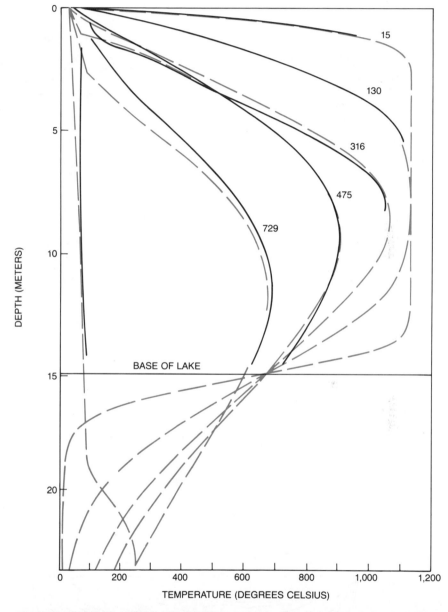

TEMPERATURE IN ALAE LAVA LAKE is shown as a function of depth. The solid black curves are the result of measurements made with steel-sheathed thermocouples inserted into drill holes over a four-year period. Each curve is labeled according to the number of days that had elapsed since the lake formed. The broken colored curves come from a computer-generated numerical model of the thermal activity in a lava lake. The model incorporates the external processes that affect the flow of heat in a lava lake (cooling by the surrounding air and rock that are initially at a temperature of about zero degrees C. and cooling by rainwater), the latent heat of basalt (the heat released by crystallization) and variations in the density, porosity and thermal conductivity of the lava. The average difference between the calculated temperature and the measured temperature is less than two degrees. Such close agreement tends to confirm the validity of the model, the main import of which is that the thermal conductivity varies with the porosity and temperature of the lava and that the vaporization of rainwater from the hot solidified crust of a lava lake has a large effect on decreasing the temperature of the lake.

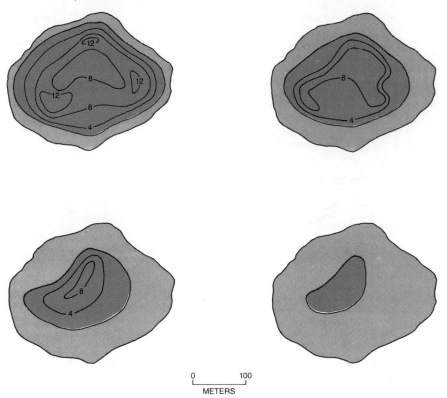

0   100
METERS

SURFACE OF ALAE LAVA LAKE is pushed up by gas bubbles that form in the underlying solidifying lava. A lava lake, like a lake of water, is shallower at the edges than it is at the center, so that the top and bottom crusts fuse at the edges (*black*), separating the molten lens from the rock enclosing the lake basin. The surface of the lake above the lens is colored. As the molten lens cools, gas in the melt is driven out of solution and pushes up the colored areas of the lake's surface. The contours indicate the rate of uplift of the surface in units of 10^{-2} centimeter per day. At the edges of the lake, where there was no solidifying lens because the top and bottom crusts were joined, the surface (*black*) subsided as the cooling lava became denser owing to thermal contraction; since this lava had already solidified, no bubbles were being generated that would make it buoyant. The contour diagram at the top left represents the lava lake from 34 to 104 days after it formed, the diagram at the top right from 236 to 293 days, the diagram at the bottom left from 293 to 345 days and the diagram at the bottom right from 345 to 400 days.

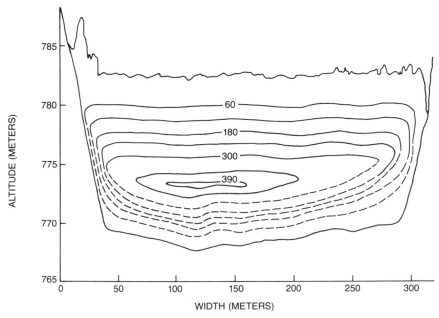

LONGITUDINAL CROSS SECTION OF ALAE LAVA LAKE shows how the isotherm, or surface of equal temperature, at 1,000 degrees C. changed as the lens of molten lava shrank. Each position of the isotherm is labeled by the number of days since the time the lake formed.

tial techniques for finding the position, size and shape of deeply buried magma bodies. Our investigation of the lake is entirely experimental, directed at filling specific gaps in knowledge and not at harnessing a particular Kilauean lava lake as a source of power.

In 1976 two holes were drilled through the crust of the lake to the top of the lens of molten lava. The properties of the lens were determined directly from core samples and from instruments lowered into the drill holes. We also tried to indirectly determine the same properties by remote techniques, such as seismic and electromagnetic measurements made at the surface of the lake; such techniques could in principle be applied to deeply buried magma inaccessible to direct drilling. An analysis of the results suggests that the seismic and electromagnetic techniques can determine the top and lateral margins of the lens of molten lava but cannot accurately estimate its thickness. It seems that these techniques perceived as melt not the entire lens of molten lava but only the hotter central part.

Working with seismometers, Bernard Chouet of the Massachusetts Institute of Technology and John L. Colp of Sandia Laboratories have mapped the approximate borders of the lens of the melt at Kilauea Iki lava lake. The seismic energy is provided by tiny earthquakes caused by the opening of thermal-contraction cracks in the regions with steep thermal gradients at the top and lateral margins of the lens; there are as many as 200 such shocks per hour. Electromagnetic studies of the lake done by Lennart A. Anderson and Charles J. Zablocki of the U.S. Geological Survey reveal approximately the same borders: a lens of molten lava 500 meters wide and 750 meters long.

The electromagnetic studies show variations in the electrical conductivity at different depths in the lake. The top layer of the top crust that is at temperatures below 100 degrees C. is quite conductive because the large fractures in it are filled with rainwater. Below this layer is a region of solidified basalt that at temperatures above 100 degrees is highly resistive because of the absence of liquid water and because the potentially conductive ions in the basalt are frozen in minerals. At depths below 40 meters lies the lens of molten lava, which is an excellent conductor because ions are free to move in it.

This year six new holes were drilled in Kilauea Iki lava lake. The lens of partially molten lava is now about 30 meters thick and consists of many crystals. It will solidify completely within five years at the most. Until Kilauea or some other volcano gives rise to a new lava lake, this unique opportunity to study magma in situ will soon be a thing of the past.

Tephra

by Laurence R. Kittleman
December 1979

*Airborne fragments from an erupting volcano, known
collectively as tephra, come in a wide variety of sizes,
shapes and compositions. The study of tephra deposits
assists in the dating of ancient events*

The word tephra, from the Greek τέφρα, meaning "ash," has come into use among geologists to describe the assortment of fragments, ranging from blocks of material to dust, that is ejected into the air during a volcanic eruption. It was first used in this modern sense a few decades ago by Sigurdur Thórarinsson, a volcanologist at the University of Iceland. Thórarinsson also coined the word tephrochronology, the dating of geological and other events by reference to their position in a sequence of tephra deposits. The two words express the essence of the current state of understanding of such matters: A volcano produces successive showers of tephra that fall throughout the surrounding countryside, forming layers that constitute a tephrochronological record of the volcano's activity.

Most tephra deposits result from volcanic eruptions in which molten rock containing dissolved gas rises in a conduit and suddenly separates into liquid and bubbles. The bubbles grow explosively, burst the surrounding liquid and give rise to a mixture of fragments and gas that is driven from the vent by the force of its own expansion and hurled far above the volcano. The fragments, which cool and solidify during their flight, are caught by winds blowing across the eruption cloud and are carried leeward, falling to the ground as much as thousands of kilometers away. The tephra falls into whatever environment happens to lie below the volcanic plume: hills, valleys, oceans, lakes, stream terraces, bogs or human settlements. There it is likely to form a persistent layer that has latent in its properties clues that can be deciphered long afterward to determine the nature of the eruption, the meteorological conditions prevailing at the time and perhaps the year or even the season of the event. Tephra influences the environments into which it falls, generating effects whose consequences often can be discerned in the stratigraphic record. Since tephra layers are formed quickly throughout a large area, they serve as handy time markers. The layers often contain or are closely associated with materials that can be dated by various means. Once a layer has been dated that date is applicable wherever the layer can be recognized, and it can be used to establish the time of any event that can be related to the tephra layer.

Tephra is found all over the world in deposits of every geological age. Its diverse properties, effects and scientific applications have attracted the attention of anthropologists, archaeologists, astronomers, botanists, chemists, climatologists, geographers, geologists, historians, oceanographers, sociologists, soil scientists and zoologists. To sample this diversity I shall summarize the current understanding of the production and dispersal of tephra and describe some ways to study the material and exploit its properties; I shall also give a few examples of eruptions that have been notable subjects of research. The discussion will mainly concern tephras of the last ice age and thereafter, for which the evidence is in some ways easiest to interpret, but it should be understood that the principles invoked apply also to older tephras.

Volcanism is the release at the earth's surface of magma, a complex molten mixture of silicon, metals, oxygen, hydrogen, sulfur and other elements, some of which are given off as gases as the magma comes near the surface. Although the nature and the origin of magmas have been studied for more than a century, several important matters remain unexplained. There is now, however, agreement on the main features.

The most likely source of the heat responsible for magma is the radioactive decay of elements dispersed in the earth's mantle, the thick layer of material that lies between the metallic core and the rocky crust. Magma probably forms in the uppermost few hundred kilometers of the mantle, but just how it is melted and mobilized is not understood. What is known is that the material of the mantle somehow melts and reaches the surface as magma at temperatures of approximately 1,100 degrees Celsius.

Magma that has spilled out onto the surface is called lava. It can have various compositions, ranging from mafic lava, which is rich in magnesium, iron and calcium and poor in silica (silicon dioxide), to silicic lava, which is rich in silica, sodium and potassium. This variability is responsible for the great variety evident in the products of volcanic activity.

Nearly every type of volcano produces some tephra. For example, comparatively quiet outpourings of very fluid lava are accompanied occasionally by fountains of molten material, whose spray solidifies into a kind of tephra. Tephra produced in this way is usually small in quantity and confined to the neighborhood of the vent. Some eruptions, called hydromagmatic, occur when water from an external source (ground water, a lake or an ocean) gains access to magma in a conduit, giving rise to violent explosions that are not necessarily accompanied by much new magma; almost all the tephra produced comes from the walls of the conduit or from shattered parts of the volcanic crater. The largest amount of tephra comes from eruptions that are accompanied by the rise of new magma. In an eruption of this type little or no lava flows out at the surface; instead the magma is converted into tephra.

The amount and the character of the tephra ejected during a particular eruption are determined largely by the properties of the magma, the most important of which are the amount of dissolved gas and the viscosity. Gas-rich, viscous magmas tend to be associated with eruptions that produce much tephra, whereas gas-poor, fluid magmas are likely to yield flows accompanied by little tephra. The viscosity of magmas is governed largely by the chemical behavior of silicon, the second most abundant element in magma (after oxygen). Mafic basaltic magmas, for example, are about 45 percent silica, and their viscosity at 900 degrees C. is approximately 10,000 poises, some 100,000 times greater than the viscosity of motor oil at room temperature. Silicic rhyolitic magmas are about 72 percent silica, and their viscosity at 900 degrees C. is close to a trillion poises. A viscosity of such magnitude surpasses ordinary experience, but the viscous behavior of silicic magmas seems to be an important factor in tephra eruptions. After the destructive eruption of Mount Pelée on Martinique in 1902, for example, a stiff protrusion of lava grew from the crater to a height of 300 meters, like a gigantic squeeze of toothpaste.

The investigation of the products of volcanic eruptions has yielded ideas that explain generally how tephra is made, although many details remain to be explored. As silicic magma rises in its conduit it cools and the pressure on it decreases. These changes cause the viscosity to increase and enable dissolved gases, mainly steam, to form bubbles. The two effects act together progressively, so that bubbles are growing most rapidly and forcefully at the same time that the enclosing liquid is becoming more viscous. Eventually the gas pressure ruptures the films of liquid between ad-jacent bubbles, and the froth disintegrates into a cloud of fragments and gas. If the final events happen near the orifice of the eruption, a mass of disrupted froth may spill out and flow down the slopes of the volcano. If the final events happen some distance below the orifice, the mixture, confined in the conduit as though it were in the barrel of a gun, is propelled upward by the steadily expanding gas and is ejected from the vent, often with enormous violence. The cloud of tephra and gas rises high above

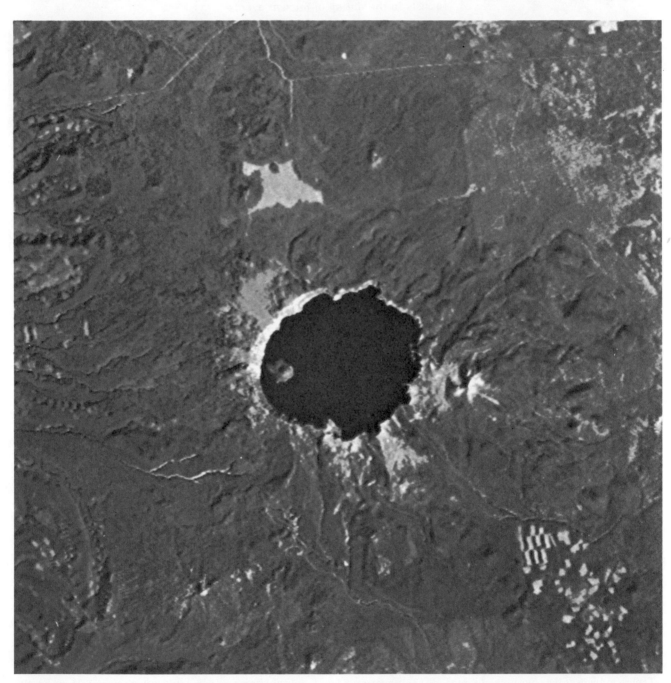

CRATER LAKE, at the crest of the Cascade Range in southwestern Oregon, is seen from an altitude of more than 900 kilometers in this false-color composite image made with data recorded by the Landsat II satellite. The lake, which is nine kilometers across and 600 meters deep, occupies a basin where the top of a volcano, Mount Mazama, once stood. The volcano erupted violently some 7,000 years ago, scat-tering about 30 cubic kilometers of tephra fragments over an area of nearly a million square kilometers. The withdrawal of so much magma left the volcano's summit unsupported and it collapsed, creating the basin (known as a caldera) that now contains the lake. The island visible near the western edge of the lake is Wizard Island, a basaltic cinder cone that rose long after the formation of the caldera.

the volcano, and particles in it are carried downwind, producing a rain of tephra that forms a deposit called a tephra mantle. To be sure, there are modes of eruption intermediate between these extremes, and there may be a continuous range of modes consistent with this general scheme. Almost innumerable variations on the process, representing intricate interactions among the amount and the composition of the magma, the rate of ascent and the rates of change of the viscosity and the pressure, account for the diversity of tephra; it is this diversity that makes it possible to distinguish the deposits of one eruption from those of another and to learn the nature of eruptions long past.

The mechanism described above accounts at least for the two main kinds of tephra eruption customarily recognized by geologists: pyroclastic flows and tephra falls. Pyroclastic flows are coherent, mobile streams of tephra and gas that usually travel along or close to the ground and may follow existing stream valleys. Such flows can extend for tens of kilometers and can travel at more than 100 kilometers per hour. The deposits formed from them are called ignimbrites. Deposits of this type are common and may cover thousands of square kilometers. In the past few decades geologists have characterized a particular kind of ignimbrite, called welded ash-flow tuff, among rocks of Cenozoic age (from 65 million years ago to the present). These deposits are the result of pyroclastic flows in which the tephra fragments have been welded together by retained heat after the flow came to rest, producing a dense rock made of compressed, deformed tephra fragments.

There have been a number of pyroclastic flows in historic times. They are dangerous, owing to their sudden onset, speed and extent. The 1902 eruption of Mount Pelée was a pyroclastic flow: a horizontal jet of tephra, gas and scalding mud that destroyed St. Pierre, a city of 30,000.

A tephra fall, the second kind of tephra eruption, is a shower of fragments borne by winds from the eruption cloud above the volcano. It is tephra of this kind that is commonly called "volcanic ash" and that forms the widespread, voluminous tephra deposits of great eruptions, some of which extend over hundreds of thousands of square kilometers. These tephra mantles will be the main subject of the rest of this article, although in general the description of their composition and mineralogical properties is applicable to other kinds of tephra as well.

Up to now I have written in general terms of the creation of tephra fragments. What are these fragments? To answer the question I must return briefly to the subject of magma. If mag-

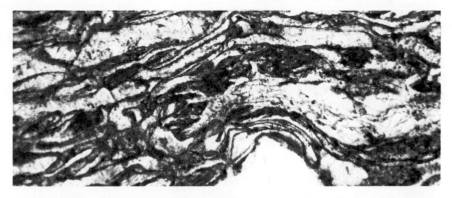

COMPRESSED, DEFORMED TEPHRA FRAGMENTS are viewed by transmitted light in this color micrograph of a thin section of welded ash-flow tuff, a variety of ignimbrite. Deposits of ignimbrite are produced by a particular kind of volcanic eruption called a pyroclastic flow, in which coherent streams of tephra and gas travel close to the ground. In the micrograph the individual tephra fragments are delineated by dark lines that were probably formed by the oxidation of iron at their surfaces. The shards were compressed in the vertical direction by the force of gravity after the pyroclastic flow came to rest but before the tephra layer cooled and hardened. The shards are bent around the edge of a large colorless grain of feldspar near the bottom. The sample, from the Dinner Creek tuff in southeastern Oregon, dates from the Miocene epoch (between five and 22 million years ago). Field of view is about a millimeter across.

MINERAL GRAINS from the tephra mantle of ancient Mount Mazama were recovered in the Blue Mountains of Oregon some 300 kilometers northeast of Crater Lake. The gray oblong object near the center of this transmitted-light micrograph is a grain of hypersthene, an iron-magnesium silicate whose orthorhombic crystal form is evident in the shape of the grain. The smaller black grain at the right is magnetite, an iron oxide, and the colorless grains at the upper left are the feldspar mineral plagioclase, a sodium-calcium aluminosilicate. All the grains have patches of glass clinging to them, remnants of the molten material in which they were suspended just before the eruption. The hypersthene grain is about 350 micrometers long.

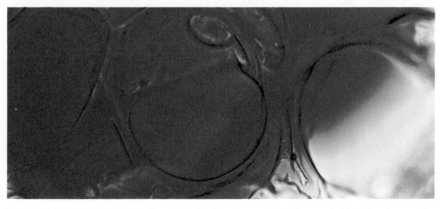

BUBBLE-WALL TEXTURE characteristic of tephra particles is revealed in this micrograph, made in transmitted polarized light, of the surface of a mineral grain from the Mount Mazama tephra. The surface retains the impressions of tiny gas bubbles present in the molten glass in which the grain was suspended prior to the volcanic eruption. Most of the glassy jacket surrounding the grain was torn away in the course of the eruption, leaving the grain covered with a thin film of glass bearing the marks of the bubbles. Bubble imprint at center is approximately 70 micrometers wide. All three of the micrographs on this page were made by the author.

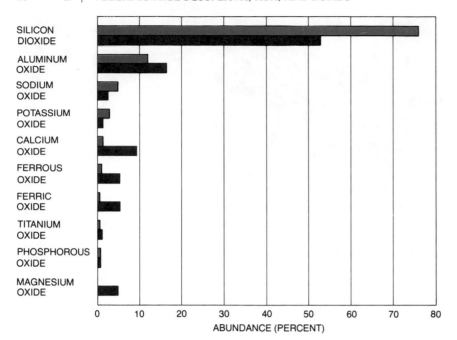

MAJOR ELEMENTS IN TEPHRAS, expressed here as oxides, exhibit a reciprocal relation between the amount of silica (silicon dioxide) present and the abundances of oxides of other elements. For example, the silicic tephra of Mount Katmai in Alaska (*colored bars*) is rich in silica but comparatively poor in oxides of calcium, iron (both ferrous and ferric forms) and magnesium. In contrast, the mafic tephra of the volcano Semeru in Indonesia (*black bars*) is poorer in silica but comparatively rich in oxides of aluminum, sodium and calcium. In general volcanic rocks exhibit broad chemical similarities among specimens with the same silica content; hence distinguishing features must sometimes be sought among the details (*see illustration below*).

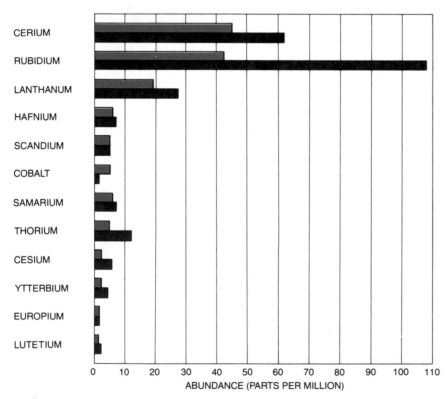

TRACE ELEMENTS IN TEPHRAS are usually present in parts per million, but their importance transcends their small abundance. They supply clues to the origin and evolution of magmas, they can often be used to date geological events and they facilitate the identification of individual tephras. Shown here are trace-element abundances found in tephras from two ancient volcanoes in Oregon: Mount Mazama (*colored bars*) and Newberry Crater (*black bars*).

ma is allowed to cool slowly, crystalline chemical compounds—minerals—grow in the liquid, drawing their substance from elements in the magma. If the growth continues to completion, the result is rock consisting entirely of an assemblage of minerals whose aggregate composition is the same as that of the magma, excluding any substances that may have escaped as gas. For most tephras, however, the process of crystallization does not go to completion. Usually magma rising in a conduit already contains some mineral grains that have grown during the ascent. The frothy liquid in which these grains are suspended is quenched suddenly when the mixture erupts. The product is a glass mixed with mineral grains and with fragments of lava or other kinds of rock broken off the walls of the conduit or the crater. In short, tephra is a mixture of glassy, mineralic and rocky fragments combined in various proportions, depending on the characteristics of the eruption that produced them. The material is called respectively vitric, crystal or lithic tephra, depending on which component is predominant. Glassy fragments are usually the main product (and sometimes the only product), but in certain tephras mineral grains or rock fragments are most abundant.

Glassy fragments are often broken walls of burst bubbles, called bubble-wall shards. These microscopic particles can be filmy plates without ornamentation, ribbed fragments, multipoint forms, crescent-shaped slivers, threads, bundles of minute tubes, strings or clusters of bubbles and so on almost without limit. Often they consist of clear, colorless glass whose curves, angles and facets sparkle brilliantly when they are suitably illuminated under the microscope, and usually they are accompanied by brightly colored mineral grains, some with a crystalline form.

Glassy fragments larger than a few millimeters are usually present in the form of pumice, a frothy rock crowded with tiny cavities that once were gas bubbles. The cavities may be roughly spherical, irregularly shaped or drawn out into long, thin tubes like the bubbles in taffy. Often the material has a distinctive silky appearance, and usually it is light enough to float on water.

There are other kinds of glassy fragments. Grant H. Heiken, a geologist at the Los Alamos Scientific Laboratory, has classified the shapes of tephra particles, relying on both transmitted-light microscopy and scanning electron microscopy. He examined both glassy and rocky particles, since there is no definite distinction between them but rather a gradation from entirely glassy fragments at one extreme to entirely rocky ones at the other. Besides bubble-wall shards he identified forms that include fragments of droplets, angular particles with or without tiny bubbles and fragments of froth that might be called micropumice. Heiken found a relation between the style of the eruption and the forms of the fragments. For example, tephra eruptions that arise from mafic magmas of low viscosity are likely to produce droplets and fragments of droplets composed of the basaltic glass sideromelane, whereas hydromagmatic eruptions tend to produce angular, chunky rock fragments with curved faces. It is possible from such observations to learn the mode of an ancient eruption by examining the forms of the tephra particles it produced.

Tephras may contain any of the many minerals that are found in volcanic rocks, but the number of common minerals is small. A single tephra ordinarily will contain only a few minerals, and it is usual for three or four kinds to constitute 80 or 90 percent of those present. Particular assemblages of minerals tend to be regional in occurrence, so that vol-

canoes of a chain or a province characteristically produce certain assemblages and rarely produce others.

Minerals of the feldspar group are common, often predominant. Included are the plagioclase feldspars (sodium-calcium aluminosilicates) and the alkali feldspars (potassium-sodium aluminosilicates). Among the ferromagnesian minerals, those with important amounts of iron and magnesium, are the pyroxenes (iron-magnesium silicates and aluminosilicates), the amphiboles (hydrous iron-magnesium aluminosilicates) and various opaque minerals (principally iron-titanium oxides). Olivine (iron-magnesium silicate), zircon (zirconium silicate) and the mica mineral biotite (hydrous potassium-magnesium-iron aluminosilicate) are sometimes present too. The proportions of the minerals and their individual optical properties vary, and it is this variation that helps to distinguish one tephra from another.

Since the mineral grains in tephra were once suspended in molten materi-

al, films of liquid have clung to them, later becoming jackets of glass that bear the impressions of tiny adjacent bubbles. These impressions create a pattern on the surface of the grain that has been named bubble-wall texture by Richard V. Fisher of the University of California at Santa Barbara. Bubble-wall texture clearly marks tephra particles that have found their way into other kinds of sediment. The glassy jackets quickly wear off during fluvial transportation, however, and so their preservation indicates that the particles have not traveled far.

Volcanic rocks generally contain about 75 chemical elements in significant quantities, of which some 10 major elements have abundances of .1 percent or more. The remainder are trace elements. Some elements not represented in analyses of rocks are erupted in important quantities as gases, chiefly hydrogen and oxygen in the form of steam, carbon in the form of carbon dioxide, sulfur in the form of hydrogen sulfide and sulfur dioxide, and chlorine and fluorine as atomic gases. The major elements, expressed as oxides, account for more than 99 percent of the material analyzed, and four of them, silicon, aluminum, calcium and sodium, account for more than 80 percent. The abundances of these elements vary reciprocally, so that rocks rich in silica are also fairly rich in sodium and potassium, whereas those poor in silica are richer in magnesium, calcium and iron. Such relations impart a broad uniformity to volcanic rocks of similar silica content, and so distinctions must be sought among the details.

The trace elements are usually present in amounts conveniently expressed as parts per million, but their importance exceeds that expressed by abundance figures alone, because they furnish clues to the origin and evolution of magmas. Some, for example, are nuclides of radioactive-decay series whose abundances can be exploited to measure the time elapsed since the tephra was formed. Relative abundances of trace elements can distinguish one tephra from another, probably with greater confidence than that provided by the major elements.

The gradation of grain sizes in tephras is large, extending from more than a meter to a few micrometers, a range of more than a millionfold. A single sample might contain fragments whose size ranges roughly from 10 centimeters to 10 micrometers, although some tephras are predominantly coarse-grained and others are fine. In 1961 Fisher proposed a grain-size classification that is now widely adopted: particles up to two millimeters in diameter are called ash, those between two and 64 millimeters lapilli and those more than 64 millimeters blocks or bombs. Blocks are fragments that were solid when they were erupted, whereas bombs were molten.

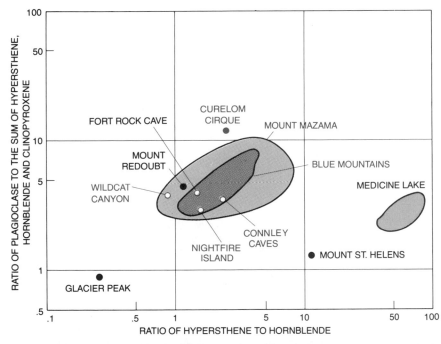

RELATIVE ABUNDANCES OF MINERALS can be analyzed in order to distinguish individual tephra deposits from one another. In this diagram, for example, the ratio of the abundance of plagioclase to the sum of the abundances of hypersthene, hornblende and clinopyroxene is plotted with respect to the abundance ratio of hypersthene to hornblende for a variety of tephra samples. The dots and the open circles designate single analyses of tephra samples; the shaded areas encompass the results of a number of analyses of the same tephra layer. The samples represented by the black dots were obtained from Mount Redoubt in Alaska and from Glacier Peak and Mount St. Helens in Washington; those represented by the light shaded areas were obtained from Mount Mazama in Oregon (*light colored area*) and Medicine Lake Highlands in California (*light gray area*). The open circles stand for analyses of tephra layers at archaeological sites at Wildcat Canyon, Fort Rock Cave and Connley Caves in Oregon and at Nightfire Island in California, all of which can be shown on the basis of this and other evidence to be tephra deposits from Mount Mazama. A tephra layer in Curelom Cirque bog in northwestern Utah (*colored dot*) is also recognizable as originating from the Mount Mazama eruption, as is a tephra layer in the Blue Mountains (*dark colored area*). Although tephras from Mount Redoubt and Mount Mazama resemble each other mineralogically, their only connection is that they come from volcanoes of the same chain. The possibility that they could be confused is ruled out because they are far apart and show different abundances of trace elements.

Volcanic bombs are still soft enough to change shape during their flight and to flatten or spatter when they land. The various types of volcanic bomb are named, for example, cannonball, spindle, bread crust, cow dung, ribbon or fusiform, depending on their shape or the appearance of their surface. Occasionally they will explode after landing, owing to the expansion of gas in a molten interior under a solid crust.

It is important to realize that technically the term volcanic ash means tephra of a particular grain size, a definition that may conflict with ordinary usage. The generic word tephra applies to fragments of any grain size.

Tephra fragments created near the orifice of a volcano are driven upward by the escaping gases. Observations at the beginnings of eruptions show that the eruption cloud rises rapidly. Thórarinsson noted during the eruption of Mount Hekla in Iceland in 1947 that the cloud rose at an average rate of about 70 meters per second during the earliest stages. The eruption cloud above Mount Asama in Japan in 1936 was observed by Takeshi Minakami of the Earth-

quake Research Institute in Tokyo to rise at a rate of more than 100 meters per second. Clearly the individual particles in such clouds must also rise rapidly. The velocity at which the particles are ejected (their muzzle velocity, so to speak) is not necessarily the same as that of the fluid propellant, however, because solids in the stream of gas tend to sink gravitationally. The ejection velocities of the fragments can be estimated by measuring the distance they were tossed. For example, Minakami calculated that during the eruption of Mount Asama in 1783 particles five centimeters in diameter were carried to an altitude of nearly 15 kilometers, implying an ejection velocity of at least 500 meters per second.

C. A. Wood and F. M. Dakin of the University of Addis Ababa, studying a hydromagmatic crater in Ethiopia named Ara Shatan, calculated that pieces about 40 centimeters in diameter had an ejection velocity of 90 meters per second and those about a meter across an ejection velocity of nearly 70 meters per second. They estimated that the stream of gas had a velocity of almost 125 meters per second, which would have called for a pressure of perhaps

200 atmospheres. William G. Melson of the Smithsonian Institution and his associates concluded from studies of the eruption of the Central American volcano Arenal in 1968 that the ejection velocity was as much as 600 meters per second under a pressure of about 4,700 atmospheres.

Tephra-laden eruption clouds rise as high as 80 kilometers, smaller fragments being carried to the highest altitudes owing to their greater ejection velocity and their greater susceptibility to the lift of the rising gas. Wind catches the particles, which are carried leeward at the velocity of the wind. Then they begin to fall, ultimately reaching a terminal velocity. The path followed by each particle is governed by the resultant of the wind velocity and the terminal velocity; a particle will strike the surface at whatever distance it can travel with the wind in the time needed for it to fall to the ground from its starting altitude at the terminal velocity. The smallest particles of course fall farthest from the volcano.

The terminal velocity is that steady speed reached by a solid particle falling freely in a fluid at which the force of gravity, acting downward, equals the force of fluid drag, acting upward. The idea, long a fundamental of fluid mechanics, is essential to studies of grain size in tephra and other sediments. The terminal velocity can be calculated with the aid of a formula that takes into account the gravitational constant, the coefficient of fluid drag, the density of the solid particle, the density of the fluid (in this case air) and the diameter of the particle. The numerical solution of this seemingly simple formula is difficult, however, because the coefficient of drag itself depends on the velocity, the shape of the particle and the viscosity of the fluid. Fluid properties such as temperature, density and viscosity, all of which affect the terminal velocity, vary with altitude, so that the terminal velocity will actually change during a particle's fall. G. P. L. Walker and his colleagues at the Imperial College of Science and Technology in London analyzed the matter in detail and also found experimentally that better results can be obtained if the particles are treated mathematically as though they were cylinders rather than spheres. (The particles do in fact have irregular shapes.) The wind velocity also may vary with altitude, influencing the path of the falling particle independently of its rate of fall.

Surveys of tephra deposits reveal the relations between grain size and distance of transport. My own work with the tephra of Mount Mazama in southern Oregon (the present site of Crater Lake) confirms the underlying dependence on terminal velocity. The average grain size of tephra deposits measured at sites arrayed along a line radial to the source decreases systematically with in-

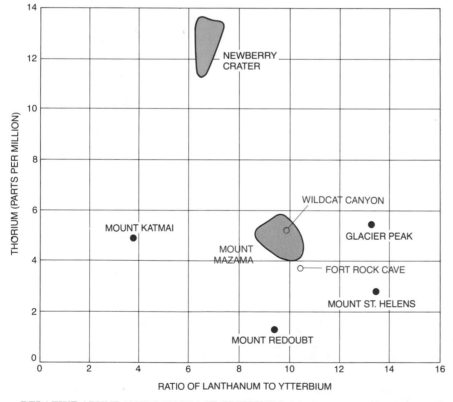

RELATIVE ABUNDANCES OF TRACE ELEMENTS can also serve to characterize individual tephra deposits. In this case the abundance of thorium is plotted with respect to the abundance ratio of lanthanum to ytterbium for a variety of tephra samples. As in the diagram on page 61, the dots and the open circles designate single analyses of tephra samples, whereas the shaded areas encompass the results of a number of analyses of the same tephra layer. The tephras of Newberry Crater and Mount Mazama in Oregon, Mount Katmai and Mount Redoubt in Alaska and Glacier Peak and Mount St. Helens in Washington are clearly distinguished. The locations of open circles, representing tephra layers at archaeological sites at Fort Rock Cave and Wildcat Canyon in Oregon, show that both samples are from Mount Mazama.

creasing distance. Work on Mount Mazama tephra by Howel Williams of the University of California at Berkeley and Gordon G. Goles of the University of Oregon and on the tephra of Mount Hekla by Thórarinsson shows also that the thickness of the deposits decreases regularly outward, depending on the distance from the source and other factors peculiar to each deposit.

Tephra dispersed according to these processes forms around the volcano a tephra mantle whose shape, volume and variations in grain size are governed not only by the characteristics of the eruption but also by the factors described above. Typically the mantle has a teardrop shape, with the volcano near the narrow end. The mantle will be longer for strong winds than for weak ones, and variable winds or different wind directions at different altitudes may give rise to a multilobed form. It is significant that some characteristics of the mantle have little or nothing to do with happenings at the volcano but express only variable meteorological conditions.

Tephra can be carried for great distances. Tephra from the 1947 eruption of Mount Hekla, for example, fell in Helsinki, 3,800 kilometers away, carried there at an average speed of about 75 kilometers per hour. Tephra from the eruption of Mount Mazama some 7,000 years ago can be recognized in western Saskatchewan, nearly 2,000 kilometers from the volcano. The finest tephra reaches into the stratosphere, where it can be carried around the world by strong global winds.

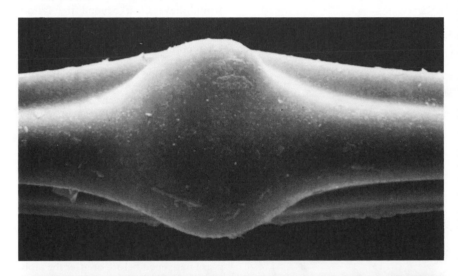

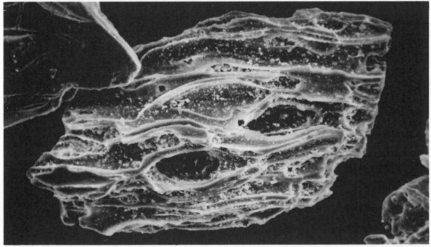

TWO UNUSUAL TEPHRA PARTICLES appear in these scanning electron micrographs made by Grant H. Heiken of the Los Alamos Scientific Laboratory. The glassy threadlike particle extending across the upper micrograph was recovered from the vicinity of Mauna Ulu volcano in Hawaii. Tephra particles of this kind, called Pele's hair, are characteristically produced in lava fountains consisting of low-viscosity basaltic magma. The prominent bulge in the glass envelope, which measures about 35 micrometers in diameter, contains an olivine crystal. The fragment of pumice in the lower micrograph is from the 1883 eruption of Krakatoa in Indonesia. This blocky particle of silica-rich glass had many small gas bubbles in it when it was ejected from the volcano, as can be seen from the characteristic array of elongated hollow structures that make up much of the bulk of this kind of tephra particle. The pumice fragment is about 250 micrometers long. The dustlike specks that adhere to it are smaller fragments of glass.

Tephra eruptions are brief. The greatest production of tephra is usually confined to a short, acute event, although a period of intermittent lesser activity may last for months or years from the first signs to the waning episodes. Nearly all the tephra from the eruption of the Indonesian volcano Krakatoa in 1883 fell in two days, although the eruption lasted for 90 days. The 1947 eruption of Mount Hekla lasted for 180 days, but 86 percent of the tephra fell on the first day. Prehistoric eruptions were doubtless similar. Peter J. Mehringer, Jr., and his associates at Washington State University studied pollen associated with a layer of tephra believed to have come from Mount Mazama in a bog in western Montana. They deduced from the estimated rates of the influx of pollen and from the indications of seasons in the relative abundances of the different kinds of pollen that the Mount Mazama eruption began in the fall and may have lasted for three years. Michael T. Ledbetter of the University of Rhode Island, working with a technique based on the terminal velocities of tephra particles in the ocean, estimated that a prehistoric tephra layer he studied was deposited within 21 days.

Samples from long sequences of tephra layers can be obtained in some geological settings, particularly deep-sea sediments, which can be sampled by modern submarine-drilling techniques. For example, Ken F. Scheidegger and Laverne D. Kulm of Oregon State University studied tephra layers interbedded with other sediments in the Gulf of Alaska, the oldest of which were deposited about eight million years ago. The tephras almost certainly came from the Aleutian Islands, a chain of volcanoes believed to lie above a subduction zone, where according to the modern theory of plate tectonics oceanic crust is plunging into the mantle and being consumed. Chemical analyses show cyclical variations in silica content: silicic eruptions are manifested in samples of tephra originating in modern times, 2.5 million years ago and five million years ago; less silicic ones are manifested about a million years ago, 3.5 million years ago and eight million years ago. The causes of this cyclicity are not known, but Scheidegger and Kulm surmise that the variation may reflect changes in the rate of the subduction of crust into the Aleutian Trench.

The theory of plate tectonics holds that most volcanism arises at the active margins of plates, either along subduction zones or along the lines where new ocean floor is created by the upward flow of material from the mantle and plates move away from each other. Volcanoes are regarded as indicators of tectonic activity, not only at present but also throughout the past; the geological record of volcanism is considered to be

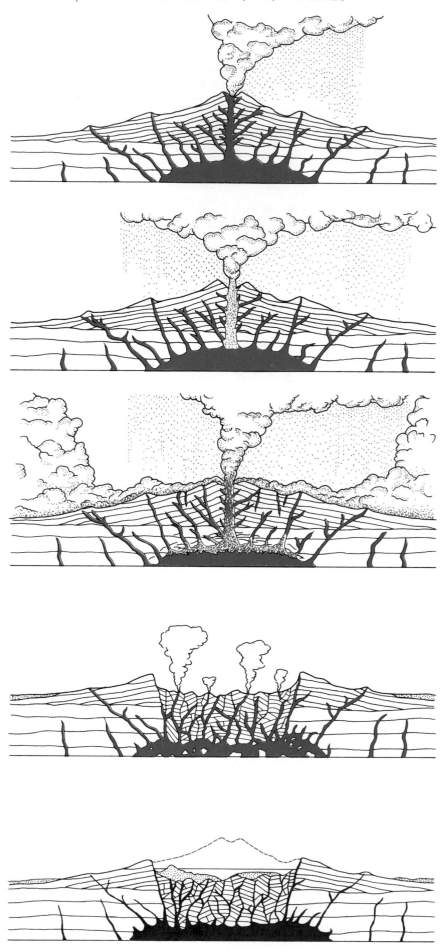

a record of the motions of plates. The ideas of plate tectonics influence the interpretation of geological data in surprising ways. For example, Dragoslav Ninkovich and William L. Donn of the Lamont-Doherty Geological Observatory argue that the motions of the plates must be taken into account in interpreting sequences of tephra layers in deep-sea sediments. They propose that the motion of a plate would over a period of time carry a sampling site from beyond the fallout range of a tephra source into that range, so that earlier parts of the record would lack tephra layers. If the site were assumed to have been stationary with respect to the source, then the record would be interpreted, in this view erroneously, to show no evidence of early volcanism when in fact that evidence would be lacking only because the site would have been beyond the fallout range at those earlier times.

A tephra mantle forms quickly everywhere throughout its extent. It represents the time of the eruption that created it wherever it is found, and this remains the case even if parts of the mantle later are removed by erosion or buried by younger deposits. In applying the principles of tephrochronology two things are essential. First, a tephra layer must be recognized and distinguished from all other layers with which it might be confused; second, the age of the layer must be determined. If these tasks are accomplished, then any deposit or object that can be associated with the tephra layer is thereby dated. Some such associations may only establish that an event happened before or after the tephra fall, but others are more exact, as when an object is enclosed within a tephra layer or between two layers. It is helpful to know the identity of the volcano that produced the layer, but it is not strictly necessary.

Relative dating based on established sequences of geological strata, assemblages of fossils, pottery types, historical events and the like has long been an important part of geological research and remains so. In recent years, however, physical and chemical means of dating also have become essential; they depend little on judgment, and their re-

FORMATION OF CRATER LAKE following the great tephra eruption of Mount Mazama in about 5000 B.C. is shown in this series of drawings, based on the work of Howel Williams of the University of California at Berkeley. An estimated 30 cubic kilometers of volcanic material was ejected from the volcano in the course of the eruption, depleting the magma chamber under it. The summit then collapsed into the empty chamber, creating the vast caldera, which subsequently filled with rainwater and melted snow to form the present lake. A smaller eruption in about the 10th century A.D. led to the emergence of Wizard Island from the floor of the caldera.

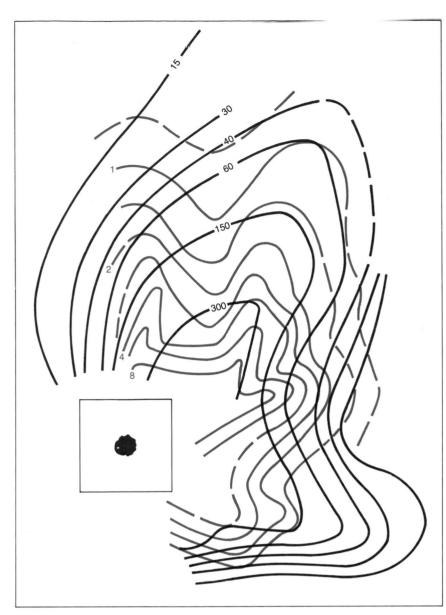

CONTOUR MAPS of the tephra mantle resulting from the eruption of Mount Mazama trace lines of equal thickness in centimeters (*black*) and lines of equal median grain size in millimeters (*color*). The different shapes of the two sets of contours demonstrate the general rule in volcanology that the direction in which the greatest volume of tephra is carried need not be the same as the direction of travel of the largest particles. In this case, for example, the absence of a detectable eastern lobe in the grain-size contour map suggests (among other possible explanations) that a westerly wind at a high altitude might have carried a large volume of comparatively fine-grained tephra particles eastward, dispersing them over a much larger area. The overall shape of the mantle in both maps indicates that the main winds during the tephra fall were from the southwest. The map of equal-thickness contours was originally drawn by Williams; the map of equal-grain-size contours was prepared by Richard V. Fisher of the University of California at Santa Barbara. The square outline at lower left shows the extent of the Landsat image reproduced on page 58; Crater Lake is drawn to scale at the center of the square.

sults can be expressed in years rather than in less precise relative terms. All these dating methods are based on the measurement of properties that change regularly and predictably with the passage of time. The main ones are carbon-14 dating, potassium-argon dating, fission-track dating, obsidian-hydration dating and amino acid–racemization analysis. These techniques are costly and are dependent on the analysis of materials that are not always available;

tephrochronology can extend the significance of the few analyses needed to date a tephra layer.

Tephras can be recognized individually by comparing their properties. Techniques for this purpose have been developed by, among others, Ray E. Wilcox of the U.S. Geological Survey and by Virginia C. Steen-McIntyre of Colorado State University. In such studies it is assumed that samples of tephra from one mantle resemble one another

more closely than samples from different mantles do. Among the properties that can be examined are the mineralogical composition, the physical properties of the individual minerals, the grain size, the thickness, the color and the chemical composition of the entire tephra or any of its constituents. In general, characterizations based on more than one property are more reliable. A typical chemical study might include the analysis of as many as 30 trace elements, although comparisons among three or four elements might be adequate, particularly in combination with tests of other properties.

Collaborating with me, Keith Randle and Goles conducted a neutron-activation analysis of trace elements to characterize some tephras in the Pacific Northwest. The measured abundance of thorium and the abundance ratios of lanthanum to ytterbium served to distinguish among the tephras we examined. The tests showed that tephra layers found at certain archaeological sites came from Mount Mazama. Mineral abundances that I studied in the same tephras also varied from layer to layer and confirmed the source of the tephra layers at the archaeological sites. These findings were consistent with archaeological evidence of the antiquity of artifacts under the tephra. One of the archaeological sites was Fort Rock Cave, which was excavated by Luther S. Cressman and later by Steven F. Bedwell, both of the University of Oregon. Under the tephra Cressman found sandals made of sagebrush bark, one of which was dated by carbon-14 analysis and found to be some 9,000 years old.

An earlier tephrochronological study by Thórarinsson revealed 15 tephra layers from the eruptions of Mount Hekla, which he dated on the basis of archaeological and historical evidence. The earliest layer fell in A.D. 1104, two centuries after the settlement of Iceland; the latest fell in 1970. The eruptions of Mount Hekla have come at intervals that average 61 years; the shortest interval was 15 years and the longest was 120 years. The longer the volcano is in repose, the more silicic the tephra of the ensuing eruption is, so that long quiet intervals are likely to be followed by a large eruption.

Dwight R. Crandell, Donal R. Mullineaux, Wilcox and their associates at the U.S. Geological Survey have documented more than 25 tephra layers from Mount Rainier, Mount St. Helens and Mount Mazama in deposits in and near Mount Rainier National Park in Washington; all were shown by carbon-14 dating to have been deposited within the past 20,000 years. The layers have been useful in correlating and dating the glacial deposits with which they are interbedded. One layer is interbedded with the deposits of a great flood in the Columbia River basin resulting from the sudden release of a glacial lake through the failure of an ice dam. The carbon-14 date of 13,000 years ago, previously obtained for the tephra layer, also dates the flood.

Deposits dating from the last ice age and earlier in the Great Plains region of the U.S. have been correlated with the help of tephra-layer evidence by John D. Boellstorff of the University of Nebraska. He recognized some 20 tephra layers that range in age from about 11 million years to 400,000 years. The tephra layers are interbedded among fluvial, windblown and glacial deposits. They were dated mainly by the fission-track method, and they can be distinguished from one another by their chemical characteristics, particularly their abundances of manganese, iron and samarium. Some of the tephras came from as far away as northeastern California and Yellowstone National Park in Wyoming.

William F. Ruddiman and L. K. Glover of the U.S. Naval Oceanographic Office found tephra particles dispersed in submarine sediments about 9,300 years old in the North Atlantic. The tephra particles are too coarse to have been carried by wind from the volcanoes nearest the sampling sites. Ruddiman and Glover believe the tephra was blown westward from volcanoes in Iceland and fell on pack ice drifting through the Denmark Strait. From thence the ice was carried southeastward by ocean currents into the North Atlantic, where it melted and dropped its burden of tephra on the ocean floor. The data are believed to indicate the existence of ocean currents flowing southward between Greenland and Iceland about 9,300 years ago. Such a deduction can serve in turn to aid in the reconstruction of the climatic patterns that may have prevailed at the time.

When tephra is dispersed over a large area, its effects spread with it. Consequences propagate through ecosystems, alter landscapes, nurture soils and influence climate. They are brief, prolonged or delayed, subtle or conspicuous. Some aspects are as yet poorly understood, and research is only beginning on means of recognizing consequences in the historical record.

In the case of large eruptions the effects near the volcano are anything but subtle. The land is buried under tens of meters of tephra, and the vegetation is destroyed. Farther away, however, the tephra is not hot enough when it falls to start fires, and at the margin of the tephra mantle the accumulation may be only a few millimeters thick.

The responses of plants are varied. Trees may be damaged by the weight of clinging tephra, and shorter vegetation is often smothered. The larger and more mobile animals may not be harmed, but they may move out when their food is gone. The response of animals is not easy to predict. A geologist watching the eruption of Paricutín in Mexico in 1943 told of his group's dismay when an unexpected burst of activity caught them near the crater; the dog that was with them, however, was curled up asleep in the warm tephra on the ground.

Freshly fallen tephra can give off substances that are poisonous in sufficient concentration, mainly carbon dioxide, fluorine, chlorine and compounds of sulfur. After the eruption of Laki in Iceland in 1783 nearly 80 percent of the sheep in the country died of fluorine poisoning, which they got from eating contaminated grass. Measurements made after the 1970 eruption of Mount Hekla showed as much as 4,000 parts per million of fluorine in dry grass; 250 parts per million would kill livestock in a few days.

Studies by William I. Rose, Jr., of the National Center for Atmospheric Research and by Paul S. Taylor and Richard E. Stoiber of Dartmouth College indicate that tephra particles may scavenge substances from the eruption cloud. These investigators leached freshly fallen tephra in water and analyzed the solutions they obtained. Apparently sulfur, chlorine and fluorine in the eruption cloud quickly form acidic aerosols that adhere to tephra particles and react with them to form salts that are dissolved by rain. The tephra fall from the volcano Cerro Negro in Nicaragua in 1968, for example, had a volume of about 13 million cubic meters, from which there may have been released into the environment, among other elements, some 8,900 tons of chlorine, 140 tons of fluorine and 2.1 tons of copper.

In spite of the apparent devastation caused by tephra falls, recovery can be rapid. After the initial prolonged eruption of Paricutín oak trees survived where the tephra was less than 1.5 meters deep, and grass lived to send out new shoots where it was buried by as much as 25 centimeters of tephra. The eruption of Mount Katmai in Alaska in 1912 deposited 25 centimeters of tephra on Kodiak Island, where low vegetation was smothered and the branches of trees were weighted down. By September plants sprouted from crevices in the compacted tephra. Most of the trees survived and replaced their damaged foliage, and by 1915 grasses and other low vegetation were thriving. Icelanders have traditionally resumed their normal routine after a volcanic eruption by clearing tephra from fields and around buildings; today they are aided by machines.

Not all the effects of tephra are harmful. Repeated tephra falls in the Tropics probably renew the fertility of the soil, which otherwise would be quickly leached of nutrients in the prevailing climate. Generally tephra seems to have a mulching effect, improving moisture-holding capacity and enhancing fertility. Studies of Mount Mazama tephra by Frederick W. Chichester of Oregon State and by Ronald R. Tidball of the University of California at Berkeley indicate that intricately shaped glassy particles hold water through capillarity, thereby prolonging storage and keeping moisture in contact with the glass, which easily decomposes chemically, releasing nutrients. The Mount Mazama tephra, in which there are few obvious signs of soil formation, supports forests of ponderosa pine and lodgepole pine.

After prolonged weathering and burial glassy fragments change to clay minerals, mainly montmorillonite, and most of the glassy tephras of Miocene age (between five and 22 million years old) or older have been partly or entirely transformed into clays. One of the most conspicuous properties of montmorillonite is its stickiness when wet. This property is familiar to anyone who has tried to drive on a wet unpaved road in terrain underlain by altered tephra, a common experience in the Western states.

Tephra the size of dust that reaches very high altitudes is carried around the world by stratospheric winds as a kind of haze, dubbed the volcanic dust veil by Hubert H. Lamb of the University of East Anglia. The dust may remain suspended for years in the stratosphere, where one of its effects is the attenuation of the solar radiation that reaches the ground. Lamb developed a method of quantitatively assessing the intensity of the dust veils produced by eruptions in historic times in a way that is uniformly applicable to modern eruptions and to earlier ones for which accurate instrumental observations are not available. Recently Bernard G. Mendonca and his colleagues at the National Oceanic and Atmospheric Administration reported daily observations of the transparency of the atmosphere in Hawaii between 1957 and 1977. Their measurements showed a significant decline in transparency during the upsurge of explosive volcanism between 1962 and 1966. Some investigators believe such changes, if they are protracted, can influence climate by lowering the average global temperature and can perhaps even contribute to the initiation of glaciation.

The mere mention of volcanoes is likely to evoke images of rivers of lava and fountains of incandescent molten rock. Nevertheless, volcanic eruptions that produced mainly tephra are among the most notorious, as the following examples will convey.

In about 5000 B.C. a volcano on the crest of the Cascade Range in southern Oregon erupted cataclysmically, spreading tephra over an area of almost a million square kilometers. The volcano was Mount Mazama, which during the last ice age grew to a height of perhaps 3,700 meters. The mountain supported glaciers from time to time, and eruptions of glassy lavas from vents on its upper slopes sometimes filled and overflowed *U*-shaped glacial valleys. Then came the tephra eruption, first a tephra fall and then a lesser pyroclastic flow, together more than 30 cubic kilometers of volcanic material. The summit of the mountain, left without the support of the magma that had been inside it, collapsed. That it collapsed rather than blew up is indicated by the absence of rocky fragments in the tephra. The collapse occurred along steep, curved fractures girdling the top of the volcano, so that today an unbroken circle of great cliffs faces a central basin that contains a lake: the centerpiece of Crater Lake National Park. In about the 10th century a small basaltic cone, Wizard Island, grew up from the floor of the depression and then stopped growing. Is the volcano now dead? I believe few volcanologists would say so.

It is known that people lived around the mountain when it erupted, because artifacts have been found both above and below the tephra layer, but research in progress has yielded only a little evidence of the effects of the tephra fall on human populations and the plants and animals that sustained them. What did witnesses to the eruption think of it? Some evidence may have survived. Ella Clark of Washington State University, a compiler of native American traditions from the Pacific Northwest, has reported a legend attributed to the Klamath people, who call the mountain Lao Yaina. The story relates that the Chief of the Below World, who lived within the mountain, was angered by the refusal of a young woman to become his wife, and he vowed to destroy her people with the Curse of Fire. He stood raging at the summit, fire spewing from his mouth, and the mountain shook and rumbled. Red-hot rocks hurtled through the sky, burning ashes fell like rain and flame devoured the forests. The people prayed to the Chief of the Above World, who caused the Chief of the Below World to be driven back inside the mountain, and the top of the mountain fell in on him. In the morning the high peak was gone. The essence of the modern interpretation is all there, in a spoken tradition that may have endured for 7,000 years, although perhaps some allowance should be made for the possibility of embellishment by later translators with some knowledge of modern geology.

In the eastern Mediterranean the volcano Thera, on an island in the Aegean Sea, erupted in about the middle of the 15th century B.C., spreading tephra southeastward over an area that includes Crete. At that time Crete was the center of the Minoan civilization, a society of maritime traders who dominated the Aegean and extended their influence throughout the eastern Mediterranean. The people had houses with several stories, courtyards and sewers. Their arts are magnificent, particularly their frescoes depicting the sport of bull vaulting. Archaeological excavations have revealed that the Minoan culture declined and disappeared suddenly after about 1450 B.C. Archaeologists have been puzzled by the unusual suddenness of the collapse, and when the approximate date of the Thera eruption became known, it was proposed that the tephra fall was the cause. The matter has been investigated by Dorothy B. Vitaliano of the U.S. Geological Survey. Data from submarine core samples analyzed by Norman D. Watkins and his associates at the University of Rhode Island indicate that as much as 10 centimeters of tephra may have fallen on eastern Crete, perhaps not enough to destroy a flourishing civilization but enough to be harmful. One carbon-14 date shows that the volcano probably erupted between 1499 and 1413 B.C., but Vitaliano found tephra from Thera in Minoan buildings that had been abandoned and covered by about 1500 B.C., according to the dating of artifacts in the area. The mystery will not be solved until more accurate dates are available, but for now it appears that the Minoan society remained vigorous for some decades after the eruption of Thera. Vitaliano believes the tephra fall could have contributed to the decline, together with later natural and political events.

Perhaps the most famous volcanic eruption in Western history is that of Mount Vesuvius on August 24, A.D. 79. It is famous because of a tephra fall that buried the Roman city of Pompeii, preserving it intact until modern times, along with the many articles of daily life abandoned when their owners fled. Most of the 20,000 residents seem to have escaped, but 2,000 skeletons have been found, along with cavities in the tephra that are the molds of bodies. Plaster poured into the cavities makes a cast, the surface of which preserves details of the body, including hair and clothing. The victims often had their hands over their face, as though to ward off suffocating gases, and it is likely that the tephra did give off gases such as carbon dioxide and sulfur dioxide. Some of the dead had been carrying bags of coins. A number of the skeletons were found well above the bottom of the tephra layer, an indication that people survived the early part of the tephra fall only to die later.

The study of tephra for its own sake and for what it can tell about the past is rewarding enough, but there is a further reason for such an undertaking, if one is needed: people living today are not immune to events such as those that overtook the people of the Pacific Northwest, Crete and Pompeii. Tephra falls cannot be prevented, but knowledge of their characteristics can help scientists, public officials and citizens in general to forestall chaos, prevent loss of life and limit damage.

6

The Eruptions of Mount St. Helens

by Robert Decker and Barbara Decker
March 1981

The volcano's current cycle of activity is part of a pattern extending over 4,500 years. Indeed, its violent eruptions of last year were predicted by volcanologists on the scene

The violent eruption of Mount St. Helens on May 18 of last year was one of the most closely monitored, most intensively photographed and most fully documented volcanic eruptions in history. It was also the first volcanic eruption in the 48 contiguous states of the U.S. since the considerably milder eruptions of Lassen Peak from 1914 to 1917. The eruption of Mount St. Helens displaced 2.7 cubic kilometers of volcanic rock (including .5 kilometer of new magma, or liquid rock), devastating an area of more than 500 square kilometers and causing one of the largest avalanches in recorded history. In the past century the Mount St. Helens outburst was clearly surpassed in magnitude only by the eruptions of Santa Maria in Guatemala in 1902, Krakatoa in Indonesia in 1883 and Katmai in Alaska in 1912, which respectively expelled some five, six and 12 cubic kilometers of magma (with the volumes reduced to correspond to the density of solid rock).

Even Katmai is dwarfed by ancient eruptions that buried thousands of square kilometers under huge deposits of ash and rock tens to hundreds of meters thick in Japan, New Zealand, Central America, the western U.S. and many other volcanic regions of the world. The volume of material expelled by these enormous eruptions ranged from 100 cubic kilometers to more than 1,000. Did prehistoric volcanism exceed anything conceivable today, or has human experience provided too brief a perspective? Most geologists think the latter is true. The eruptions of the past century are probably only small samples of the volcanic energy still at the earth's command.

In the months since the eruption of Mount St. Helens the energy released by the eruption has often been described in terms of the energy released by nuclear explosions. The comparison is useful but somewhat misleading. Not only is the source of energy totally different; so too is the rate of energy release, or power. The thermal and mechanical energy released at Mount St. Helens on May

18, 1980, was about 1.7×10^{18} joules. Since a nuclear explosive rated at one megaton releases about 4.2×10^{15} joules, the Mount St. Helens eruption was equivalent to a 400-megaton explosion, one nearly eight times more powerful than that of the largest nuclear device ever detonated. The comparison is misleading because substantially all the energy of a nuclear explosion is transformed into thermal and mechanical energy in a flash, so that the almost instantaneous power in watts of a nuclear explosion is essentially the same as the energy in joules. (One joule for one second equals one watt.) In the Mount St. Helens eruption the 1.7×10^{18} joules was spread over nine hours, for an average power of about 5×10^{13} watts (1.7×10^{18} divided by 32,400 seconds). The volcano's sustained power output might therefore be compared to the serial detonation of some 27,000 Hiroshima-size bombs: nearly one a second for nine hours. For another comparison, the power generated by Mount St. Helens on May 18 was on the order of 100 times the generating capacity of all U.S. electric-power stations.

Mount St. Helens has erupted during at least 20 intervals over the past 4,500 years, often enough for the Indians of the American Northwest to know it as Loowit, the Lady of Fire. Before the 1980 eruptions Mount St. Helens was last active between 1831 and 1857. It is one of 15 major volcanoes in the Cascade Range, stretching northward from Lassen Peak in California to Mount Garibaldi in British Columbia. The Cascades are part of the "ring of fire," the volcanic ranges that nearly surround the Pacific. The mountains are thrust up at subduction zones where the moving tectonic plates of the Pacific Basin plunge under the confining plates to the west, the north and the east. The subduction of the Pacific plates generates great earthquakes and provides the molten rock and pressure that powers the volcanoes of the ring of fire. In historic times alone nearly 400 volcanoes have

been active on the margins of the Pacific Basin.

The level of volcanic activity seems to be governed in part by the rate at which the thrusting plates plunge under the bordering plates. In Indonesia and Japan, where the annual subduction rate is six or seven centimeters, there is usually at least one volcanic eruption per year. In the Cascades the lower incidence of eruptions appears to be related to the lower rate of convergence of the North American plate and the Juan de Fuca plate immediately to the west: two or three centimeters per year.

In the early 1960's an effort to evaluate the hazards presented by the dormant but potentially active volcanoes of the Cascade Range was undertaken by Dwight Crandell and Donal Mullineaux of the U.S. Geological Survey. They began their program with a study of volcanic deposits in the vicinity of Mount Rainier, 75 kilometers north-northeast of Mount St. Helens. Mount Rainier had last erupted sometime between 1820 and 1854. Guided by carbon-14 dates of earlier eruptions and following the geologist's dictum that what has happened before can happen again, Crandell and Mullineaux were able to forecast in a general way the potential hazards of each volcano they studied.

Crandell and Mullineaux' report on Mount St. Helens was issued in 1978. They concluded that Mount St. Helens had a bad record: over the preceding 4,500 years it had been more active and more explosive than any other volcano in the contiguous 48 states. In that period its eruptive products included lava domes too viscous to flow from their vent, large volcanic-ash falls full of lumps of pumice, flows of pyroclastic rock (hot fragments shattered and fluidized by eruptive activity), flows of lava and massive flows of mud in the stream valleys radiating from the volcano. The average interval between eruptive periods was 225 years.

On the basis of their study Crandell and Mullineaux stated: "In the future Mount St. Helens probably will erupt

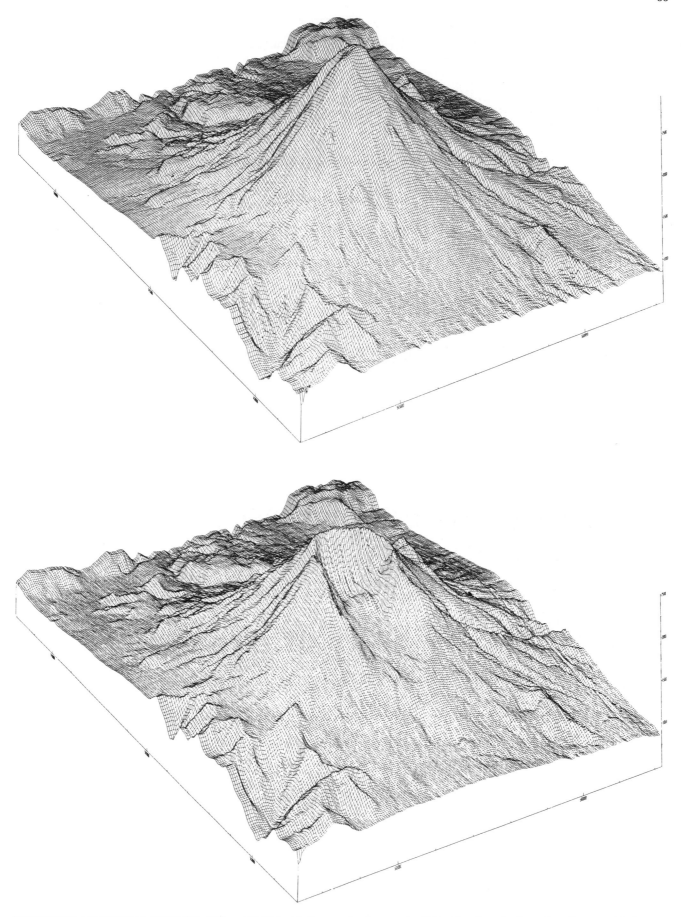

COMPUTER-GRAPHICAL MODELS of Mount St. Helens before the explosive eruption of last May 18 (*top*) and afterward (*bottom*) show how the eruption removed the top 400 meters of the mountain, leaving a crater 750 meters deep. In these views the mountain is seen from the northeast. They were created by the Digital Elevation Model program at Western Mapping Center of U.S. Geological Survey.

violently and intermittently just as it has in the recent geologic past, and these future eruptions will affect human life and health, property, agriculture and general economic welfare over a broad area.... The volcano's behavior pattern suggests that the current quiet interval will not last as long as 1,000 years; instead an eruption is more likely to occur within the next 100 years, and perhaps even before the end of this century."

Many volcanic eruptions are preceded by swarms of small earthquakes. Although not all the volcanoes of the Cascade Range are monitored by seismographs, Mount St. Helens fortu-

nately was. The University of Washington had a seismograph on the west flank of the volcano that was linked to Seattle by telemetry. On March 20, 1980, at 3:47 P.M. Pacific Standard Time an earthquake of magnitude 4 on the Richter scale occurred under the mountain. When this unusual event was followed by an increasing number of local earthquakes, it became apparent that a major earthquake swarm was in progress. In order to improve the recording and locating of the earth tremors additional seismographs were installed.

On March 25 the seismic energy released by the swarm reached its peak rate: 47 earthquakes of magnitude 3 or

more occurred within a 12-hour period at shallow depths under the mountain's north flank. The first small explosions of steam came two days later, beginning at 12:36 P.M. and forming a new crater about 70 meters across on the snow- and ice-covered summit. Large east-west cracks also developed in the snow and ice, indicating that a down-faulted block was forming across the summit area. A second crater opened on March 29. Blue flames, possibly burning hydrogen sulfide, were visible from the air at night. Ninety-three small eruptions of steam and ash were observed on March 30.

On April 1 the seismographs recorded the first volcanic tremor, a more or less

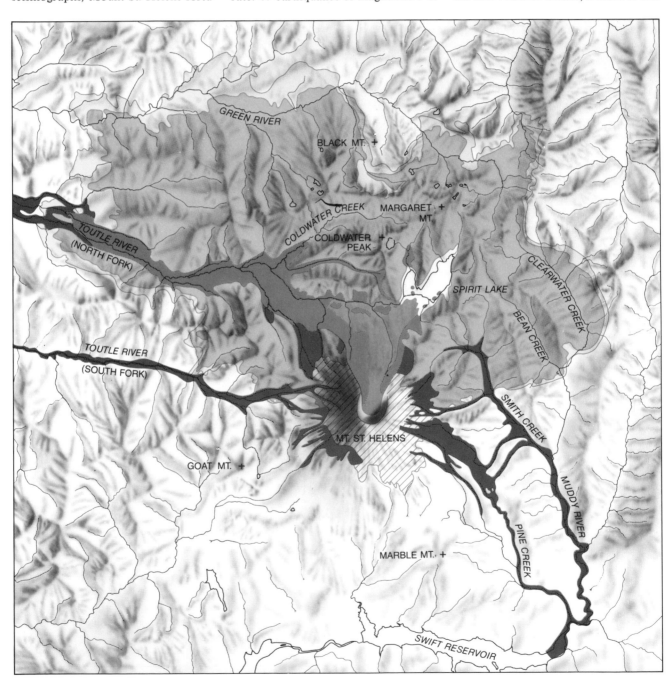

GEOLOGIC MAP outlines the region devastated by the May 18 eruption. The area in dark color is pyroclastic deposits: hot bits of fragmented rock. Hatched area is ash falls. The area in medium color is the one in which trees were blown down. The area in light color is the one in which trees were left standing but with their needles killed by heat. Light gray is the avalanche deposits; dark gray, mudflows.

continuous ground vibration observed at many active volcanoes. The precise cause of such a tremor is not known, but it presumably reflects the movement of magma or the rumbling release of gas previously dissolved in the magma.

By this time the new volcanic eruption, the first in the Cascade Range since Lassen Peak quieted down in 1917, had become a magnet for the curious. Nolan Lewis, Director of Emergency Services for Cowlitz County, Wash., in which Mount St. Helens is situated, reported that "Sunday [March 30], when the weather was clear, the roads up the mountain looked like downtown Seattle at rush hour." When the potential danger of violent eruptions became more and more apparent, Governor Dixy Lee Ray ordered additional blockades established on roads leading to the mountain.

The small steam and ash eruptions continued, some as single explosions, others as pulsating jets lasting for hours. Columns of steam and ash climbed to three kilometers above the summit. The crater of the mountain grew to a single oval basin 500 meters by 300 meters across and 200 meters deep. The ash consisted of fragments of old volcanic rock; the gas emissions included small amounts of carbon dioxide, sulfur dioxide, hydrogen sulfide and hydrogen chloride together with large volumes of steam. The small explosive eruptions apparently resulted from ground water high in the volcanic cone being heated above the subsurface boiling point and flashing suddenly into steam, much as a geyser does but with energy enough to incorporate ash particles and blast out a crater. The energy released by all the steam explosions up to May 18 is estimated at 10^{14} joules.

As the high level of seismic activity continued (about 50 earthquakes of magnitude 3 and greater per day) there was another ominous sign. Several observers had noted as early as March 27 that the down-faulting of the block across the summit of the mountain

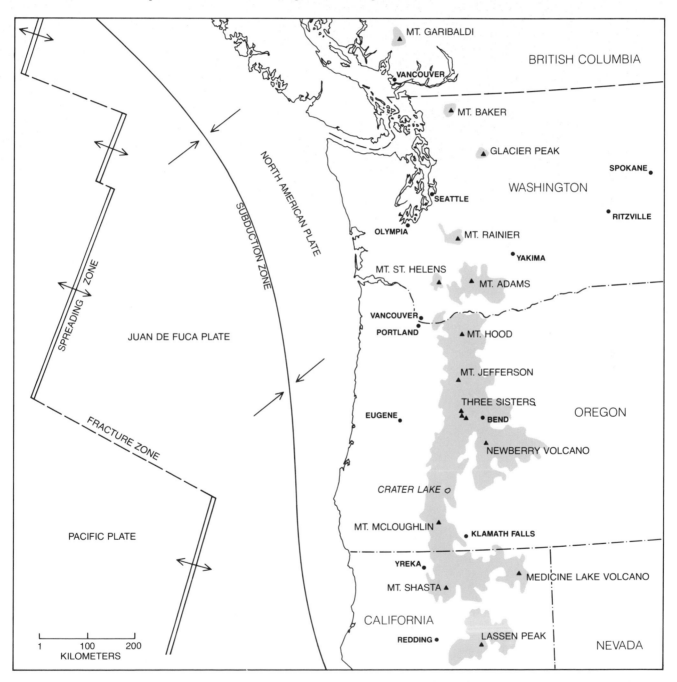

TECTONIC MAP shows the relation between the North American tectonic plate and the Juan de Fuca and Pacific plates to the west. At the subduction zone the Juan de Fuca plate is plunging under the North American plate, giving rise to the volcanoes of the Cascade Range (*small black triangles*). Colored areas are volcanic deposits less than two million years old. Data are from the U.S. Geological Survey.

seemed to be accompanied by a bulge, or uplift, of the high north flank of the volcanic cone. The bulge continued to grow early in April and was manifested by a spreading network of large cracks in the cover of ice and snow. Photogrammetric maps made by the U.S. Geological Survey from aerial photographs taken on April 12 disclosed that the bulge was nearly two kilometers in diameter and had already moved up or out by as much as 100 meters.

Ground surveys conducted in late April and early May established that the bulge was continuing to expand northward more or less horizontally at a rate of about 1.5 meters per day. The rapidly deforming area was directly over the center of the earthquake zone two kilometers below. The consensus of observers on the site was that the continuing seismic swarm and major surface deformation were good evidence that magma was being injected under the volcano at a shallow depth. They expected a major eruption or an avalanche from the expanding north face. The only questions were how soon it would come and how violent it would be.

Robert Christiansen, the investigator in charge of the U.S. Geological Survey's monitoring program, reviewed the historic sequence of activity at other volcanoes, specifically Lassen Peak, the only Cascade Range volcano with a well-observed eruption, and Bezymianny on the Kamchatka Peninsula on the Pacific coast of the U.S.S.R., which had exploded violently in 1956. Christiansen concluded that an eruption comparable to that of Lassen in 1915 was the most likely course for Mount St. Helens. An eruption on the scale of Bezymianny, however, could not be ruled out. A third possibility was that all the activity might subside without a major eruption.

Meanwhile, on May 7, after about two weeks in which there had been little visible activity, small explosions of steam and ash began again. Although seismic activity had continued unabated and the bulge had grown steadily, the lack of dramatic visible activity led residents of the region to question the political authority that was keeping the area around the mountain closed. On May 15, 16 and 17 earthquakes and bulging continued, but there was no sign of steam or ash.

At 7:00 A.M. Pacific Daylight Time on May 18 Dorothy and Keith Stoffel, two Washington geologists, boarded a light plane at Yakima airport near Mount St. Helens and set off on their first reconnaissance flight. They made several passes around and over the volcano. No activity was visible, although the morning was bright and clear. At 8:32 the mountain was shaken by an earthquake of magnitude 5.1 centered below its north flank. At that moment the Stoffels were directly over the summit and look-

ing down from a height of 400 meters. They noted several small ice falls starting on the steep sides of the crater. Fifteen seconds later they were the closest witnesses to the onset of a huge volcanic eruption triggered by one of the largest landslides of historic times.

"The whole north side of the summit crater began to move instantaneously as one gigantic mass," Dorothy Stoffel recalled. "The entire mass began to ripple and churn without moving laterally. Then the whole north side of the summit started moving to the north along a deep-seated slide plane."

Seconds later there was a vast explosion. Curiously, the Stoffels neither felt nor heard it even though they were just to the east of the summit. From their position the initial explosion cloud seemed to mushroom sideways to the north and then plunge down the slope. Their first concern was survival. Even though they dived at full throttle to gain speed, the expanding gray cloud gained on them. They finally escaped it by turning south. Behind them giant ash clouds boiled upward, thrusting north and northwest. To the east the clouds expanded into billowing mushroom shapes, illuminated by lightning bolts thousands of meters long. Half an hour later the Stoffels landed in Portland.

In the avalanche they had witnessed from the air more than two cubic kilometers of crushed rock and glacier ice plunged into Spirit Lake and the north fork of the Toutle River. Fluidized by exploding steam, the avalanche accelerated rapidly to velocities as high as 250 kilometers per hour. One lobe of the gigantic mass plowed through the west arm of Spirit Lake and northward into the valley beyond. An adjacent lobe swept across another valley with such momentum that it went over a ridge 360 meters high that bounded the valley on the north. The bulk of the fluidized debris funneled down the valley of the Toutle, forming a hummocky deposit 21 kilometers long, one to two kilometers wide and up to 150 meters deep. The gravitational energy of the avalanche was about 5×10^{16} joules (roughly the equivalent of 12 megatons).

As the great avalanche of rock and ice suddenly released the pressure within the volcanic cone, superheated ground water flashed into steam. Simultaneously dissolved gases exploded from the shallow magma body recently intruded into the upper magma core of the mountain. The steam blast, magma explosion and giant avalanche combined to form a lateral blast of hot (up to 300 degrees Celsius), dense, debris-filled steam clouds that surged northward from the breached mountainside at speeds of between 100 and 400 kilometers per hour. The steam blast and its fluidized charge of volcanic rock fragments devastated 550 square kilometers of mountain terrain north-

west, north and northeast of Mount St. Helens. The ground-hugging black clouds rolled over four major ridges and valleys, going as far as 28 kilometers from their source.

The destruction was complete. For the first few kilometers entire trees one to two meters in diameter were uprooted and swept away with the roiling explosion cloud. Beyond this was a blowdown zone 10 to 15 kilometers across, where prime Douglas firs were snapped like matchsticks. At the outer limits of destruction trees were still standing, but their needles were scorched beyond recovery.

The first impression from viewing the blowdown zone was that a great shock wave or concussion front had knocked the trees down in a pattern radial to the exploding peak. This was not, however, substantiated by the evidence. Survivors near the edge of the devastated area heard only a moderately loud explosion or roaring sound two to three minutes before the black cloud with its hot hurricane winds descended on them. The velocity of the front of the steam blast clouds was well below the speed of sound. On closer inspection the pattern of felled trees showed turbulent eddies and curving streamlines. Near the edge of the devastated area the trees tended to be blown down in down-valley directions, even where that meant they fell toward the source of the surging clouds.

Gravity apparently energized the dense, fluidized mass as the energy of the initial steam blast waned. Then as the turbulent internal winds abated, ash and rock debris that were carried in the dense clouds settled to the surface in deposits that decrease in thickness, with distance from the source, from about a meter to a centimeter. Angular fragments up to several tens of centimeters in diameter of both old volcanic rocks and fresh, hot ones were carried in the blast clouds to distances of 10 to 15 kilometers. Trees were charred and blackened on the blast side out to seven kilometers on the northwest and north, and as far as 18 kilometers on the northeast.

The fluidized character of the masses of fragments carried by the blast is even more evident in the deposits on steep slopes. Here after the initial deposition secondary flows formed ponded deposits tens of meters thick in basins and valley bottoms. The total volume of the blast deposits is about .18 cubic kilometer. Of this amount about .06 cubic kilometer is magmatic: fresh volcanic rock. The heat energy provided by the magmatic component was approximately 2×10^{17} joules.

Another probable source of energy for the steam blast explosion was the superheated ground water inside the volcano. Assuming a porosity of 15 percent in a volume of two to three cubic kilometers filled with water at an aver-

BEFORE-AND-AFTER PHOTOGRAPHS of Mount St. Helens show the extent of the new crater. The photograph at the top was made by one of the authors (Robert Decker) in June, 1970. The elevation of the summit, seen here from the north-northeast, was 2,950 meters, rising from a base with an elevation of about 1,000 meters. The photograph at the bottom was made by Ray Foster of the Sandia Laboratories in July, 1980. The crater, seen here from the north, is two kilometers across. The elevation of the rim of the crater is between 2,400 and 2,550 meters, that of the floor of the crater between 1,800 and 1,900 meters. Pyroclastic flows cover much of foreground.

age temperature of 175 degrees C., this yields an additional 10^{17} joules of energy. Steam in the explosion cloud probably was generated in two ways. Heat in the flashing ground water would generate 4.4×10^{10} kilograms of steam, and the additional heat in the .06 cubic kilometer of magmatic fragments could convert another 8.8×10^{10} kilograms of water into steam. The total of 1.3×10^{11} kilograms of steam would expand to 220 cubic kilometers at 100 degrees C.

and atmospheric pressure. The assumed mass of heated ground water inside the volcano (3.75×10^{11} kilograms) is nearly three times more than is needed to supply the calculated volume of steam. These energy and steam-volume figures are only rough estimates, yet they yield figures that are reasonable in terms of the 550-square-kilometer area devastated by the ground-hugging lateral blast clouds. The final blast deposit was a layer of wet ash up to six centimeters deep,

containing in many places the pea-size mud balls volcanologists call accretionary lapilli. These accretions formed around nuclei provided by raindrops condensing out of the steam clouds.

By 9:00 A.M. on May 18 the worst of the eruption was over, but the vertical column roared on, reaching altitudes in excess of 20 kilometers for much of the day until it began to wane at about 5:30 P.M. The source of this almost continuously exploding and uprushing column

VERTICAL BEFORE-AND-AFTER VIEWS of Mount St. Helens were made on May 1, 1980 (*left*), and June 19, 1980 (*right*), by U-2 aircraft of the National Aeronautics and Space Administration. The peak is at the lower right. The film was infrared-sensitive and the

of gas and ash was the effervescing shallow magma body being progressively cored out to greater depths. The abrasive uprush continued to enlarge the horseshoe-shaped crater that had initially been formed by the avalanche and the lateral blasts.

High-altitude winds blew all day to the northeast, and ash began to fall on towns in central Washington by mid-morning. At Yakima, 150 kilometers away, the first ash fall was a "salt and pepper" layer consisting of sand-size fragments of dark rock and of lighter-colored feldspar crystals. On top of this layer was deposited a thicker one of silt-size particles of volcanic glass. Thirty kilometers north of Yakima the ash fall was about 20 millimeters deep. To the east the fine ash was deeper, reaching more than 70 millimeters near Ritzville, Wash., 330 kilometers from Mount St. Helens. Here the texture of the ash was like talcum powder.

Near Spokane, Wash., 430 kilometers northeast of the volcano, the ash was only five millimeters deep, but by 3:00 P.M. visibility was reduced to three meters in near darkness. At about noon on May 19 a trace of ash fell on Denver. Within three days the ash cloud had crossed the U.S. The weight of the measured ash fall was equal to .15 cubic kilometer of magma and represented a dissipation of heat energy amounting to 5×10^{17} joules. Studies of the traces of

prints are in false color. The red areas are equivalent to the green of vegetation, mostly Douglas fir. Gray areas in the photograph at the right are those devastated in the eruption of May 18. Over the peak in the photograph at the right is a fume cloud issuing from the crater.

76

MOMENT OF ERUPTION on May 18 (actually 20 seconds after the eruption began) was captured by Keith and Dorothy Stoffel, geol-ogists who were flying over the mountain at the time. The eruption was preceded by beginning of the avalanche. The Stoffels escaped.

DOME OF LAVA was formed in the crater between June 13 and June 20. It was 300 meters wide and 65 meters high and was destroyed by the eruption of July 22. This aerial photograph was made by Maurice and Katia Krafft of Centre de Volcanologie at Cernay in France.

ash that fell beyond the area of measured ash thickness and the very fine ash and sulfuric acid aerosols still suspended in the stratosphere suggest that an additional volume of .1 cubic kilometer of magma was dispersed.

Judging from experience with other eruptions that have injected dust and aerosols into the stratosphere, the very fine particles will take a year or two to settle out. The effect on world climate, if any, is not yet evident.

Floods and mudflows were another major aspect of the eruption. The flows consisted of a slurry of volcanic ash and fine rock particles mixed with water, and they had the consistency of wet cement. The ash blanket near the mountain and the crushed rock in the avalanche deposit provided the solid matter; the extra water probably came from several sources: melting snow and ice, the water displaced from Spirit Lake and the north fork of the Toutle River by the avalanche deposit, the water from the breached hydrothermal system that did not flash to steam, and condensing steam.

The first mudflow crest came in the south fork of the Toutle River near Silver Lake at 10:50 A.M. on May 18. It exceeded by 30 centimeters the highest flood level of historic times. The largest mudflow came from the north fork of the Toutle and crested at about 7:00 P.M., destroying the gauge station near Silver Lake. High mud and water marks indicated that it had exceeded historic flood levels by nine meters.

Downstream from the Toutle, mudflow deposits clogged the channels of the Cowlitz River and caused severe shoaling of the navigation channel in the Columbia River. The volume of mud deposited was about .1 cubic kilometer. Roughly the same volume of water would have been involved in mobilizing the mudflows.

Sometime after the initial avalanche and steam blast eruption pyroclastic flows of fine ash and pumice blocks began rushing down the north slope of Mount St. Helens through the breach in the newly formed crater. These fluidized emulsions of hot rock and glass fragments mixed with hot volcanic gases, being denser than the ascending ash cloud, issued from the crater under the uprushing cloud. Successive flows poured down the north slope at speeds of up to 100 kilometers per hour, covering the earlier landslide and blast deposits and reaching the south edge of Spirit Lake. When these hot (300–370 degrees C.) pyroclastic flows came in contact with water, they set off secondary explosions, sending steam and ash clouds as high as two kilometers. The ash and pumice flows continued until the evening of May 18. Their total bulk volume was .25 cubic kilometer and their estimated thermal energy 3.3 × 10^17 joules.

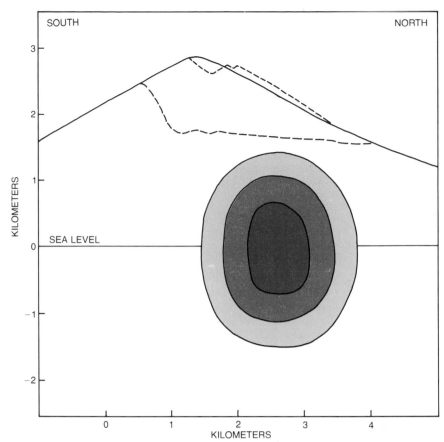

TOPOGRAPHIC PROFILES show Mount St. Helens as of August, 1979 (*solid line at top*), as of May 1, 1980, when a bulge had formed on the north slope of the mountain (*upper broken line*), and as of July 1, 1980 (*lower broken line*). The colored area at the right below the profiles is the general region of origin of the thousands of earthquakes in the swarm from March 20 through May 18. The darker the color, the higher the density of the earthquake locations. Data are from U.S. Geological Survey and Department of Geophysics of University of Washington.

The human cost of the eruption was commensurately great. Sixty-two people were killed or are missing. The economic loss, mostly to the lumber industry, was more than $1 billion. Perhaps the greatest damage was psychological. In the U.S. Northwest the Cascade Range volcanoes have been perceived as silent guardians; now they are a brooding threat.

The huge avalanche and eruption of May 18 was followed by smaller ash explosions on May 25, June 12, July 22, August 7 and October 16–18. Domes of viscous lava were extruded inside the crater on June 13–20, August 8–9 and October 18–19. The total volume of magma in the eruptions of May 25 through October 19 is about .05 cubic kilometer with an estimated energy content of 1.7 × 10^17 joules, an order of magnitude less than the May 18 eruption. The May 25 eruption occurred at night and in poor weather, but a sudden increase in the volcanic tremor at 2:28 A.M. probably marked its beginning. By 2:45 National Weather Service radar showed that the ash column had reached an altitude of 14 kilometers. It diminished in height within an hour, but the volcano continued to emit smaller ash

clouds throughout the day. New pyroclastic flows of ash and pumice blocks were also emitted, and they covered some of the same area on the north flank of Mount St. Helens swept by the earlier pyroclastic flows. Although the ash eruption was much less voluminous than the one of May 18, the wind directions were more variable and a thin layer of ash fell over wide areas of western Washington and Oregon, including the Portland metropolitan area.

The June 12 ash explosions were similar to the eruption of May 25. Volcanic tremor began in the afternoon, with the first ash emission at 7:05 P.M. reaching an altitude of four kilometers. A much larger ash eruption began at 9:09 P.M. and reached 15 kilometers. The eruption faded rapidly after midnight. Helicopter observation the next day revealed that additional pyroclastic flows two to 10 meters thick with temperatures of up to 600 degrees C. had descended toward Spirit Lake. A lava dome began to form in the explosion crater after the June 12 eruption, and when it was seen on June 15, it was 200 meters wide and 40 meters high and was covered with large, glowing cracks. The dome continued to rise about six meters per day and by June 20 had reached its greatest height of 65 meters.

It was quiet until July 22. That morning small, shallow earthquakes began under the crater area of the mountain. The earthquakes increased in number during the day, but no volcanic tremor was recorded. Suddenly at 5:14 in the afternoon of a clear summer day a large ash cloud began to roil up from the mountain. Radar registered the top of the cloud at 14 kilometers. A second ash cloud erupted at 6:25 P.M. and reached an altitude of 18 kilometers in just seven minutes 23 seconds, an average velocity of 2.2 kilometers per minute. The third and longest ash jet began at 7:01 P.M. and lasted for more than two hours, reaching a maximum height of 14 kilometers.

Although geologists and Forest Service fire fighters had left the area after being warned of the increase in small, shallow earthquakes, the eruptions were observed from helicopters and other aircraft. Pyroclastic flows poured from the vent during the second and third ash ex-plosions and spilled down the north slope of the volcano toward Spirit Lake. Richard Hoblitt of the U.S. Geological Survey gave the following eyewitness account:

"We were flying from east to west approximately one mile north of the vent as the second eruption started. Following a period of a few seconds' duration during which the rate of gas emission increased, an ash fountain was ejected to about 1,500 feet above the vent. As the projections of the fountain arced over and reached the surface in the vicinity of the vent, they gave rise to a pyroclastic flow that began to rapidly flow northward out of the amphitheater. We exited to the west as quickly as possible." Not many people have seen a pyroclastic flow from this close and lived to describe it.

These new ash and pumice-block flows were one meter to two meters thick. Their temperature was measured the next day; the maximum temperature recorded was 705 degrees C. at a depth of 1.5 meters. The ash clouds drifted to the northeast on July 22, and only minor ash falls were reported from central and eastern Washington.

A change in the pattern of gas emission had preceded the July 22 eruption. The gas emissions changed again early in August, and volcanic tremor began just after noon on August 7. Warned by these two signs, investigators in the danger areas were evacuated. An eruption began at 4:26 P.M. and rapidly generated an ash cloud that reached a height of 13 kilometers. Small pyroclastic flows swept the area below the breach on the north side of the mountain. Smaller ash eruptions continued through the late afternoon and evening, with another large outburst at 10:32 P.M. A new lava dome formed in the crater on August 8–9.

Quiet prevailed for more than two months; then small earthquakes similar to those that had preceded the July 22 eruption began on October 16. As the swarm increased an alert was issued early in the evening. It was followed by an eruption that began at 9:58 P.M. Four ash explosions over the next two days sent clouds as high as 14 kilometers and destroyed the August lava dome. Some

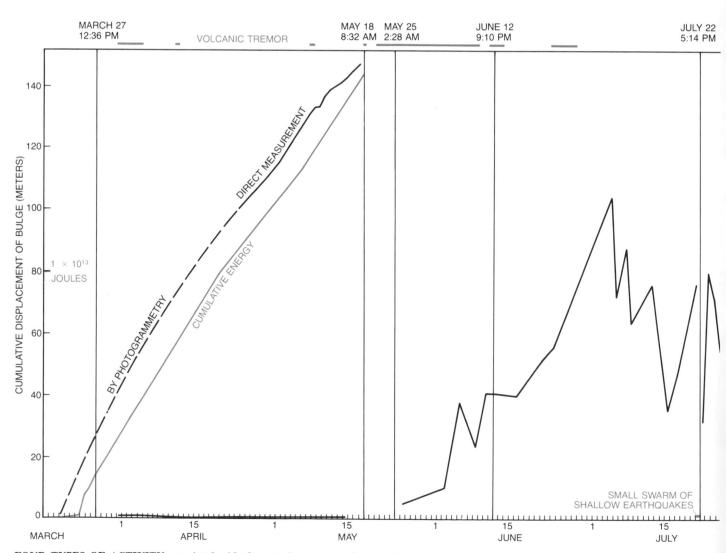

FOUR TYPES OF ACTIVITY associated with the eruptions are plotted on this graph: earthquake swarms, the formation of the bulge, the emission of sulfur dioxide gas and volcanic tremor. Important eruptions are indicated by the vertical lines. The major deformation

of the eruptions were accompanied by smaller pyroclastic flows that descended the north flank of the mountain. Light ash fell to the south and southeast, reminding the Portland metropolitan area that Mount St. Helens was still in action. A new lava dome, the largest yet to appear, arose on October 18–19.

The eruptions of volcanoes vary enormously in their explosiveness. In Hawaii fountains of incandescent lava spray spectacularly but harmlessly into the air and lava flows move slowly downslope from their vents. At the other extreme gigantic explosions destroy entire mountains when a substantial fraction of the heat energy in the magma is converted into mechanical work. One measure of the explosiveness of volcanoes is the nature of their products. Effusive eruptions are characterized by lava flows, explosive eruptions by fragmental products: volcanic ash, rock chips and blocks. In Hawaii about 98 percent of the eruption products are effusive lavas. With the volcanoes rimming the Pacific it is nearly the oppo-

site: about 90 percent of the eruption products are fragmental debris.

Volcanoes are found in three tectonic settings. The volcanoes in each of these settings differ remarkably in their explosive behavior from those in the others. As we have seen, the ring-of-fire volcanoes rise along subduction zones where tectonic plates converge. These volcanoes are generally explosive. Rift volcanoes occur where plates are diverging; their eruptions, particularly the deep submarine ones, are more effusive. Where rift volcanoes erupt in shallow water or through continental crust, however, they can be explosive. Hotspot volcanoes, which penetrate tectonic plates, are generally effusive where they occur in areas of oceanic crust (Hawaii) and explosive in areas of continental crust (Yellowstone).

The key factors governing the explosiveness of volcanoes appear to be the viscosity of the magma, the amount of gas dissolved in the magma, the quantity of ground water near the vent area and the surface pressure. Volcanic explosions are caused by the rapid expansion

of gas within the magma or in close contact with it. There is no release of energy as there is in a bomb; a better analogy is the explosion of a steam boiler. Viscosity is analogous to the strength of the boiler; the higher the viscosity, the larger the potential explosion. The availability of gas within the magma or in contact with it is comparable to the volume of the boiler. The difference between the gas pressure in the magma and the surrounding pressure is comparable to the pressure difference across the boiler walls. Steam is one of the major magmatic gases, but in volcanic explosions carbon dioxide is also important.

Consider two examples, the first a basaltic magma with a relatively low viscosity and a low content of dissolved gas. If this basalt rises through a fracture toward the surface, carbon dioxide and water begin to come out of solution as the pressure decreases. The bubbles expand against the low viscous forces and cause the foaming magma to expand. This process is most important in the last few meters before the magma reaches the surface: it gives rise to an effervescing fountain of incandescent lava at the vent. If the same basalt emerged under two kilometers of seawater, the decrease in pressure would not be sufficient for the dissolved water to come out of solution. Some carbon dioxide would probably exsolve, but the expansion of the mixture of magma and carbon dioxide would be small, and the mixture would well quietly from its submarine vent. In shallow water, depending on the rate at which the hot effusive lava mixed with the surrounding water, various degrees of explosiveness might result.

The second example is a siliceous magma with a relatively high viscosity and a high content of dissolved gas. If this dacitic magma rises toward the surface, the exsolved gas begins to form bubbles. Because of the high viscosity of the melt, however, the bubbles do not expand to their equilibrium size; they have an internal overpressure estimated to be as high as several hundred atmospheres. When enough bubbles form and the external pressure is low enough, the magma shatters into fragments that are accelerated out of a vent by the explosively released gases. The cooling of the expanding gases solidifies the magma fragments; they are ejected as hot but solid particles in an emulsion. These fluidized mixtures form eruption clouds or pyroclastic flows depending on their density and the direction and velocity of their ejection.

One of the closest analogues of the eruption of Mount St. Helens is the eruption of Bezymianny in Kamchatka in 1955–56. The Russian volcanologist G. W. Gorshkov has divided the Bezymianny eruption into five stages: (1) September 29 to October 21, 1955,

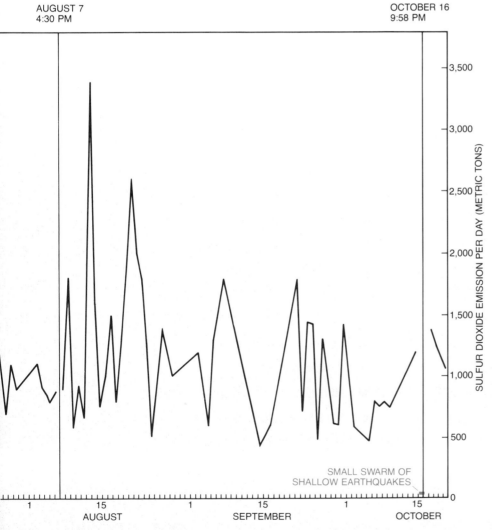

of the bulge and the earthquake energy are related to period before eruption of May 18. Data are from U.S. Geological Survey and Department of Geophysics of University of Washington.

pre-eruption earthquake swarm; (2) October 22 to November 30, 1955, strong ash eruptions; (3) December 1, 1955 to March 29, 1956, moderate ash eruptions with general uplift of the old lava dome; (4) March, 1956, gigantic explosion and lateral blast; (5) April, 1956, growth of new lava dome in the explosion crater.

The volume of rock and the area affected in the March 30, 1956, eruption of Bezymianny are similar to those of the May 18, 1980, eruption of Mount St. Helens. The major differences are that in the Bezymianny eruption the earthquake swarm preceding the first ash eruption lasted for 23 days compared with seven days at Mount St. Helens, the preliminary eruptions lasted for 160 days compared with 53 days and there were larger ash explosions in the early phase of the preliminary eruptions at Bezymianny. In the major explosion at Bezymianny most of the ejected rock was new magma, in contrast to the May 18 avalanche and explosion of Mount St. Helens, in which less than a fifth of the ejected rock was new magma. Perhaps significantly, Bezymianny has remained active since 1956. Its last major eruption occurred in 1979.

Mount St. Helens has provided a good test case for techniques of forecasting volcanic eruptions. There is no single key to success in such forecasting; all the factors must be evaluated and then interpreted in the light of geologic experience. Among the factors are the statistics of eruptions in historic times and the reconstruction of the statistics of eruptions in prehistoric times by geologic

mapping and dating. The geophysical techniques involved include monitoring, on and near active and potentially active volcanoes, seismicity, ground-surface deformation, magnetic and electrical fields, and temperatures. The geochemical techniques include monitoring the volume and composition of gases, liquids and solids emitted by volcanoes. Repeated visual observations from the ground or the air provide important data on evolving volcanic activity. Most of the world's volcanoes do not get even this basic type of examination, let alone the more sophisticated instrumental monitoring.

At Mount St. Helens some methods worked better than others. The historic-activity data indicate eruptions in the period from 1831 to 1857, but apart from identifying Mount St. Helens as an active volcano the sample is too small to have any statistical meaning. The geologic mapping and dating identified 20 eruptive periods with diverse products over the past 4,500 years and established some important patterns. Dormant intervals formed two populations: 100 to 150 years and 400 to 500 years. The last two dormant intervals before 1800 were of the short type. The eruptive products also showed that ash and pyroclastic explosions were common and that prehistoric eruptions of Mount St. Helens affected large areas. It was this kind of analysis that led Crandell and Mullineaux to make their forecast that Mount St. Helens was dangerous.

In 1980 seismic monitoring of the increasing earthquake swarm under Mount St. Helens gave a week's warning

of the first small explosive eruptions. Photogrammetry and electro-optical distance measurements documented the spectacular ground-surface deformation associated with the growing bulge on the north flank of the mountain. There was no change in the rate of seismicity or ground-surface deformation just before the May 18 eruption, but the continuation of these phenomena gave a general warning that something important was happening underground. That warning allowed Governor Ray and the U.S. Forest Service to stick to their order to keep the area closed in spite of insistent demands for freedom of access to it. Their firmness undoubtedly saved thousands of lives.

In the months since May 18, 1980, there have been successful forecasts of the smaller but still significant explosive eruptions of June 12, July 22, August 7 and October 16. Volcanic tremor preceded the June 12 and August 7 eruptions by several hours, and the unusual buildup of small earthquakes just under the mountain preceded the July 22 and October 16 eruptions by several hours. Anomalies in the pattern of the emission of gases and small deformations of the ground surface have also preceded some (although not all) of the eruptions since May 18 by hours or days.

Any change in the total monitoring pattern is suspicious, and experience helps in evaluating the change. There have been false alarms, but in any probabilistic forecasting system that seems to be inevitable. Technology has not solved the ancient problem of how sure one must be before crying wolf.

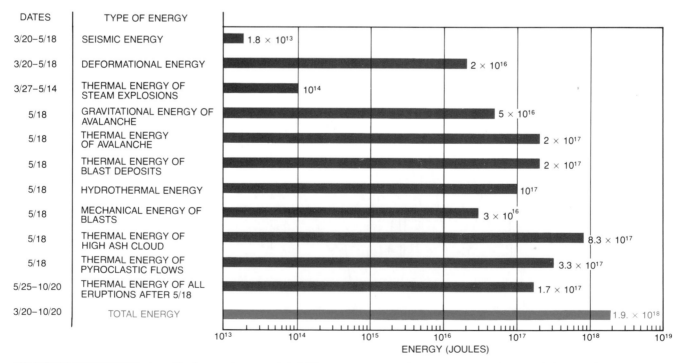

DATES	TYPE OF ENERGY	
3/20–5/18	SEISMIC ENERGY	1.8×10^{13}
3/20–5/18	DEFORMATIONAL ENERGY	2×10^{16}
3/27–5/14	THERMAL ENERGY OF STEAM EXPLOSIONS	10^{14}
5/18	GRAVITATIONAL ENERGY OF AVALANCHE	5×10^{16}
5/18	THERMAL ENERGY OF AVALANCHE	2×10^{17}
5/18	THERMAL ENERGY OF BLAST DEPOSITS	2×10^{17}
5/18	HYDROTHERMAL ENERGY	10^{17}
5/18	MECHANICAL ENERGY OF BLASTS	3×10^{16}
5/18	THERMAL ENERGY OF HIGH ASH CLOUD	8.3×10^{17}
5/18	THERMAL ENERGY OF PYROCLASTIC FLOWS	3.3×10^{17}
5/25–10/20	THERMAL ENERGY OF ALL ERUPTIONS AFTER 5/18	1.7×10^{17}
3/20–10/20	TOTAL ENERGY	1.9×10^{18}

ENERGY (JOULES)

SIX TYPES OF ENERGY released in the eruptions are plotted. The dates of the energy release are given in the column at the far left. The seismic energy, the deformational energy and the thermal energy of steam explosions were released in the period before the eruption of May 18. The scale of the bars is logarithmic. The bar at the bottom gives total energy of all these releases through October of last year.

III

VOLCANIC WINDOWS INTO THE EARTH'S INTERIOR

VOLCANIC WINDOWS INTO THE EARTH'S INTERIOR III

INTRODUCTION

Some geologists say, not without envy, that we know less about the Earth's interior than we know about outer space. In a direct sense this is true. Space probes are now moving into the outer reaches of the solar system—over one billion kilometers away. In comparison, the world's deepest drillholes reach depths of only ten kilometers, less than one one-thousandth of the Earth's diameter. However, we have a great deal of indirect knowledge about the Earth's interior, and volcanoes provide some of the best free samples. Other major sources of information are celestial mechanics, geophysics, and the study of meteorites. From these data a conceptual model of the Earth's interior can be derived. These are worth considering briefly, for the volcanic evidence gains strength when considered in the overall context of constraints from other disciplines.

The size, shape, mass, and angular momentum of the Earth have been determined in detail from a combination of surface geodetic surveys, astronomical observations, artificial satellite orbits, and gravity measurements. The calculated average Earth density is 5.5 grams per cubic centimeter, and ranges from about 2.5 grams per cubic centimeter at the Earth's solid surface to a density of about 14 grams per cubic centimeter at the 6370 kilometer-deep center.

Seismologists, monitoring and deciphering the Earth's vibrations from earthquakes and nuclear bomb tests, have divided the interior into several more or less spherical shells: crust, mantle, and core. The crust is thin beneath the oceans and thicker beneath the continents, supporting them to their higher elevations. The upper mantle contains a low-seismic-velocity, low-strength zone, perhaps composed of partially molten rock, roughly 70 to 300 kilometers below the Earth's surface. The hot, solid, but apparently plastic deeper mantle continues down to a major seismic discontinuity at 2900 kilometers—the top of the core. The outer core, down to 5150 kilometers, is molten in the sense that it does not transmit shear vibrations. The final inner core is apparently solid.

The nature of the Earth's magnetic field provides supporting evidence for a molten outer core of high electrical conductance—probably molten metal. The combined geophysical data specify the physical properties of the Earth's interior but not its compositon. That is the realm of cosmochemistry, especially the analysis of meteorites.

There are two major families of meteorites—stony and metallic. The stony ones consist largely of olivine and pyroxene, and the metallic ones generally consist of an iron-nickel alloy. Since meteorites appear to be leftover building supplies from the early construction of the inner planets, it is reasonable to

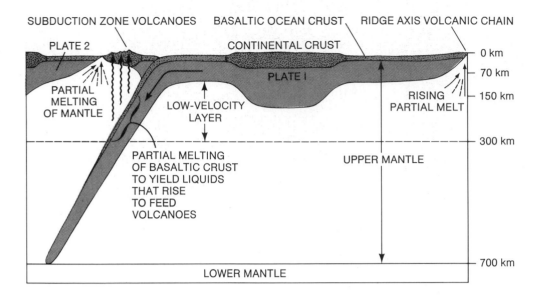

SUBDUCTION ZONE VOLCANOES   BASALTIC OCEAN CRUST   RIDGE AXIS VOLCANIC CHAIN

PLATE 2   CONTINENTAL CRUST

0 km

70 km

PARTIAL MELTING OF MANTLE

PLATE 1

150 km

RISING PARTIAL MELT

LOW-VELOCITY LAYER

300 km

PARTIAL MELTING OF BASALTIC CRUST TO YIELD LIQUIDS THAT RISE TO FEED VOLCANOES

UPPER MANTLE

700 km

LOWER MANTLE

Cross section of the Earth's upper mantle. The rigid plate is composed of solidified rock that moves on the partially molten low-velocity layer. The plates are approximately 70 kilometers thick under oceans and 100 to 150 kilometers thick under continents. The continents are parts of the plates and move with them. (After J. F. Dewey, "Plate Tectonics." Copyright © 1972 by Scientific American. All rights reserved.)

assume that the Earth's mantle has an olivine–pyroxene composition—a rock type called peridotite, and that the Earth's core is an iron–nickel alloy. The densities and seismic properties of the Earth's interior fit reasonably well with this assumption.

Once the composition and physical phases of a model Earth are specified, some calculations on its internal temperatures can be made. For the model under consideration, basalt magma temperature (1200°C) would be approached at about 100 kilometers depth, and the base of the solid mantle and top of the molten core would be at a temperature of about 3800°C. Increasing pressure would increase the melting point of iron–nickel, and the solid inner core would have temperatures in the range of 5000°C.

Volcanic rocks provide the only direct tests for models of the Earth's mantle. How do they fit with the model suggested so far? In general, quite well. The early-melting fraction of peridotite is a liquid of basalt composition. Many nodules of unmelted mantle rock carried up by ascending basalts are composed of peridotite or of closely related rock types. However, some mantle sources are quite different from others. Besides a fairly wide range in the minor elements present in volcanic rocks, basalts seem to come from two fundamentally distinct mantle sources: one that appears to have been partially melted several times, allowing it to become depleted in those elements which move easily into the melt fraction; and undifferentiated mantle that appears to have been melted for the first time. Plate tectonics provides a possible explanation for the two types of mantle rocks. Some mantle material from the low-velocity layer has apparently been recycled thorugh several generations of rifting and subduction. Hot-spot volcanoes possibly tap the deeper mantle that has not been involved in earlier generations of crustal rock formation.

In the article by O'Nions, Hamilton, and Evensen, studies of the parent–daughter ratios of long-lived radioactive elements in volcanic rocks allow even more subtle insights into the composition, structure, and evolution of the mantle source rocks from which the volcanic rocks were derived. The arguments involved must be followed with care. The Earth does not easily reveal her history and inner workings.

The final article in this series concerns Yellowstone National Park, a special place with volcanic roots that reach far into the Earth's interior. For our generation, most of Yellowstone's violence is probably in the past, but in a geologic time sense, Yellowstone seems to be only simmering. It deserves not only our admiration but also our respect.

7

The Earth's Mantle

by Peter J. Wyllie
March 1975

*This great body of hot rock accounts for 83 percent
of the volume of the earth and 67 percent of its mass.
Although the mantle is inaccessible, much has been
learned about it by indirect evidence*

The mantle of the earth is a thick shell of red-hot rock separating the earth's metallic and partly melted core from the cooler rock of the thin crust. Starting at an average depth of from 35 to 45 kilometers (22 to 28 miles) below the surface and continuing to a depth of some 2,900 kilometers, it accounts for nearly half of the earth's radius, 83 percent of its volume and 67 percent of its mass. Its influence on the crust is profound; indeed, the crust and its thin film of ocean and atmosphere are distillates of the mantle, and the driving forces that move the continents slowly about on the earth's surface arise within the mantle. Knowledge of the mantle is therefore crucial to an understanding of the structure and dynamic behavior of the earth. Notwithstanding the inaccessibility of the mantle, a considerable amount of information about it has been assembled by more or less indirect means.

The role of the mantle in influencing conditions at the surface is manifold. For example, during the 4.6 billion years since the origin of the earth in the solar nebula, melting of the more fusible constituents of the mantle has produced lavas that rise to the surface and solidify, adding new rocks to the crust and giving off the water vapor and other gases that go to the atmosphere and the oceans. On another scale gaseous carbon compounds that came from the mantle began the story of life on the earth when they provided the raw material for organic molecules.

Similarly the mantle as a driving force has multiple effects. The surface of the earth is shaped by the action of the mantle, moving very slowly below the crust. Mountains rise and persist because of this movement; without it erosion would wear them down to sea level within 100 million years or so. Movements of the mantle also cause volcanic eruptions, earthquakes and continental drift.

A cross section through the earth shows the concentric layers of the core, the mantle and the crust [*see top illustration on opposite page*]. They differ from one another in composition or physical state or both. The mantle is composed of silicate minerals rich in magnesium and iron, with an average composition corresponding to that of the rock peridotite. (The name comes from the fact that the most abundant mineral in peridotite is olivine, which is more familiar as the transparent green gemstone called peridot.) The mantle is solid but within the relatively thin zone ranging from about 100 kilometers below the surface to about 250 kilometers the rock may be partially melted, with thin films of liquid distributed between the mineral grains. This zone is called the low-velocity layer for reasons that will become apparent below.

The density of the mantle increases with depth from about 3.5 grams per cubic centimeter near the surface to about 5.5 grams near the core. It does not increase smoothly; the curve of density displays distinct steps [*see illustration on page 88*]. They indicate significant changes in the mantle rocks at depths near 400 and 650 kilometers. The distribution of density provides the basic for the calculation that the mantle makes up 67 percent of the mass of the earth.

Plate Tectonics

This static picture of a concentric, layered earth is modified by the theory of plate tectonics, which deals with the movement of lithospheric plates. The lithosphere includes the crust and part of the upper mantle and is distinguished from the asthenosphere below it by the fact that it is cooler and therefore rigid. The theory of plate tectonics provides a dynamic picture of a mobile mantle, with plates of lithosphere about 100 kilometers thick moving laterally over the asthenosphere. The surface of the earth is covered by a few large lithospheric plates and several smaller ones. These shell-like plates move with respect to one another, and the geological activity represented by earthquakes and volcanoes is concentrated along the plate boundaries.

Plate boundaries include divergent and convergent types. Below the crest of an ocean ridge material from the asthenosphere rises, melting as it moves and thus producing lava that is erupted in the central rift valley of the ridge to produce new crust. Convective movements in the mantle cause the plates to diverge from one another as new lithosphere is generated.

At convergent boundaries plates may collide, pushing up crust to form mountains, or one plate may move under another, carrying lithospheric material back into the mantle. The growth of new lithosphere is therefore balanced by the destruction of lithosphere elsewhere. Such boundaries are associated with ocean trenches and with lines of volcanoes, including arcs of volcanic islands and the volcanoes in active mountain ranges.

Many of the physical properties of the earth as a whole have been determined. The planet's size, shape and mass have been measured with precision. Its known volume and mass give a mean density of 5.5 grams per cubic centimeter, which is much higher than the density of the rocks making up the accessible crust. Therefore a significant part of the interior must be composed of material

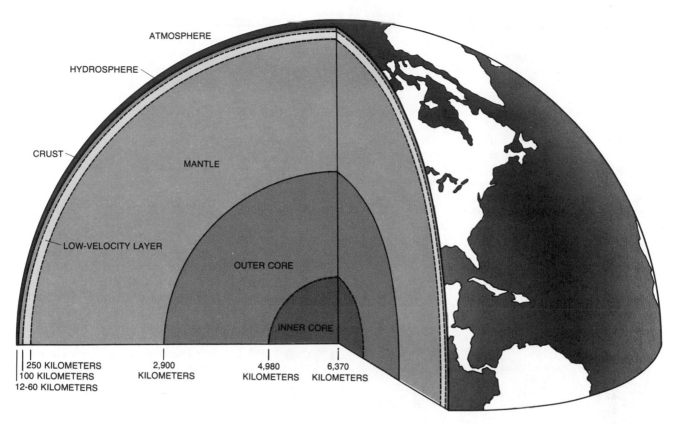

CONFIGURATION OF THE EARTH is portrayed by means of layers alone, without regard to the active processes that go on in the interior. The rocks of the thin crust are cool and rigid. Mantle rock, which is hot, is capable of slow movement. Evidence from earthquake waves indicates that the outer core consists of molten metal. Hydrosphere consists of surface and atmospheric waters.

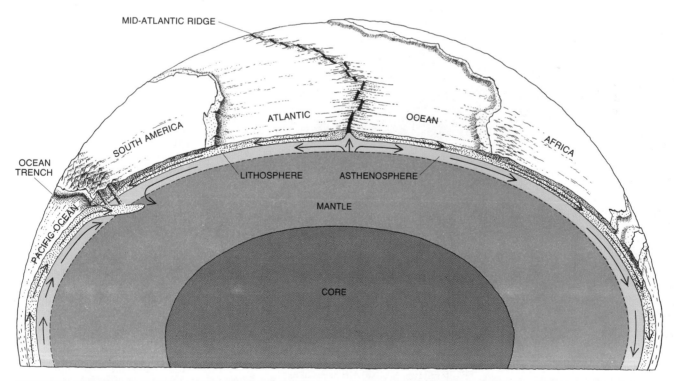

DYNAMIC EARTH is depicted in cross section as it is envisioned in the theory of plate tectonics. Plates of the lithosphere, which includes the crust and part of the upper mantle, migrate laterally over the asthenosphere, which is a hot and perhaps partly molten layer of the mantle. Material from the asthenosphere rises below the crest of an ocean ridge, melting to produce lava that is erupted to form new crust in the ocean floor. The lithospheric plates diverge as new lithosphere is generated from the rising material. The growth of new lithosphere is balanced by the destruction of an equivalent amount of lithosphere at convergent plate boundaries, where the lithospheric layer moves down into the mantle. These boundaries are associated with ocean trenches and lines of volcanoes.

with a density higher than 5.5 grams per cubic centimeter. Studies of the earth's gravitational field and the physical properties of the rotating sphere indicate that the mass is concentrated toward the center.

Evidence on the Mantle

The main body of information about the physics of the earth comes from studies of earthquake waves, which provide the equivalent of X-ray pictures of the interior. The earth rings like a bell when it is shaken by a major earthquake.

The vibrations of a bell depend on its shape and physical properties; similarly the vibrations of the earth as recorded by sensitive instruments can be interpreted in terms of the properties of the earth.

The release of energy at the focus of an earthquake produces several types of wave. The primary, or P, waves and the slower secondary, or S, waves pass through the interior of the earth. The abbreviations also serve as reminders that the energy is transmitted along the path of a ray by different phenomena: the P waves are compressional, or push-

pull, waves and the S waves are shear, or "shake," waves [see illustration on opposite page]. P waves can be transmitted through both solids and liquids; S waves can be transmitted only through materials that can support shear stresses, that is, materials that can be deformed or bent. An S wave cannot be transmitted through a liquid, because liquids cannot sustain shear; they flow too easily.

If the earth were composed of material with uniform properties throughout, the waves from the focus of an earthquake would follow straight lines, and each type of wave would travel at a constant velocity. The times taken for P and S waves to reach a particular recording station at a known distance from the focus would give the velocity of each type of wave. In actuality, however, the results obtained from many earthquakes show that the waves travel faster through the earth than would be predicted from their known velocities in surface rocks. The results show in addition that the waves that travel the greatest distance have also traveled faster. These findings mean that the velocity of earthquake waves is greater at depth than it is near the surface and also that it increases progressively with depth.

From these observations and others it is known that the waves from earthquakes are refracted and reflected within the earth. Refraction causes them to follow paths that are concave upward. Rays are reflected at levels where there is a distinct change in physical properties between layers. Reflection therefore occurs at boundaries: between the crust and the mantle, the mantle and the core and the inner and the outer core. This wave pattern demonstrates the concentric structure of the earth. S waves follow paths that reach as deep as the coremantle boundary, but they do not pass through the core. This finding is evidence that at least the outer part of the core is liquid.

Measurement of the times of travel of the waves over various paths in the mantle provides a means of calculating the velocity at which the material at each depth within the mantle transmits waves. The wave-velocity profiles show a progressive increase with depth, but the increases occur in a series of steps down to about 1,000 kilometers. These observations indicate that the upper mantle has a layered structure.

The velocity of both P and S waves decreases in the upper mantle within a layer lying from approximately 100 to 250 kilometers below the surface. This

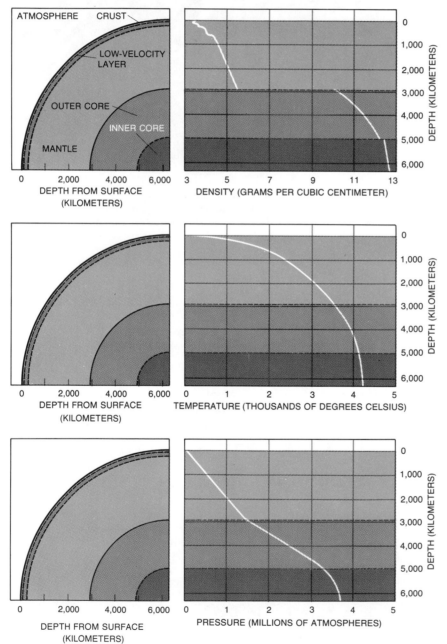

PHYSICAL PROPERTIES of the earth vary with depth from the surface. The depths on the graphs of density, temperature and pressure correspond to the depths on the cross-section diagrams at left. One atmosphere is the air pressure at sea level, 14.7 pounds per square inch.

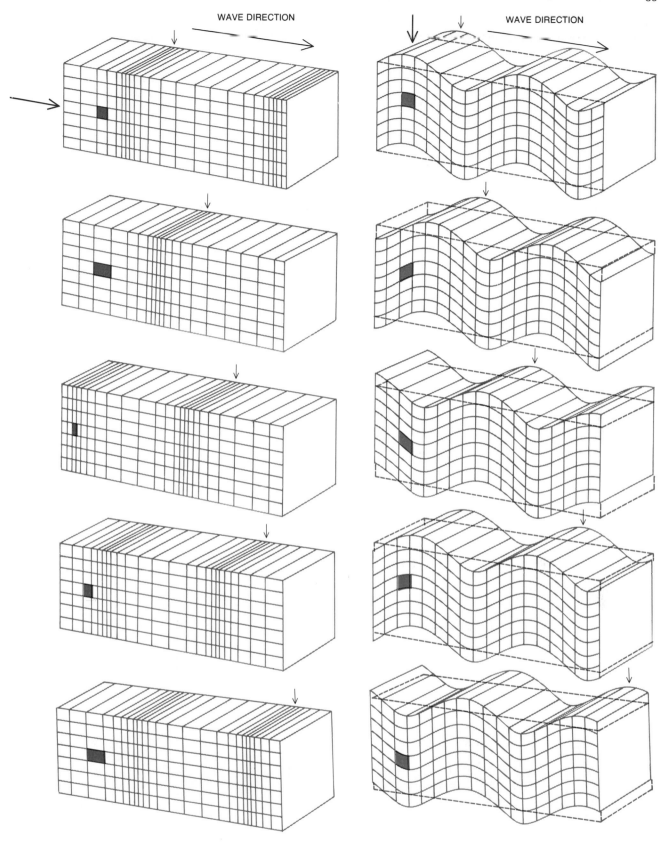

WAVE DIRECTION

WAVE DIRECTION

EARTHQUAKE WAVES of two types pass through the interior of the earth from the focus of an earthquake, where energy is released. They are portrayed here passing through a block of rock. At left is a compressional, or *P*, wave, which is started by a sudden push or pull in the direction of the wave path. The action compresses the rock, and the nearby particles move forward. They then rebound to their former positions and beyond, continuing to vibrate in this way for some time. Each small volume of matter (*color*) contracts and expands as the crest of compression moves through the rock. A shear, or *S*, wave (*right*) is started by pressure at a right angle to the wave path. Particles of rock vibrate up and down. A small piece of matter (*color*) undergoes shear deformation. The wave crest transmits energy. The illustration is adapted from *The Heart of the Earth*, by O. M. Phillips (Freeman, Cooper & Company).

is the low-velocity layer that I have mentioned. It is considered to be equivalent to the asthenosphere of the plate-tectonic model. The layer's physical properties, deduced from the wave-velocity profile and other geophysical evidence, are consistent with the presence of an interstitial liquid, which is generally thought to be a molten fraction of the rock in the layer.

Plate Boundaries

Earthquakes provide not only this static picture of the concentrically layered earth but also some aspects of the dynamic picture of plate tectonics. Since earthquakes occur only in rocks that are cool and rigid enough to fracture, the distribution of earthquake foci delineates the boundaries of the stable plates. The depths of the foci along the boundaries associated with ocean ridges are less than 100 kilometers, indicating that the mantle below the lithosphere is hot.

At the convergent boundaries where one plate sinks below another and moves into the mantle the foci of earthquakes are at depths as great as 700 kilometers. The distribution of these deep foci provides a means of mapping the layers of sinking lithosphere. Detailed studies of

wave velocities in these regions are consistent with the existence of slabs of cooler lithosphere extending to considerable depths within the mantle.

The information provided by earthquake waves about the structure and physical properties of the mantle is fairly direct. In contrast, the effort to obtain information about the chemical properties of the deep interior requires resorting to mostly indirect evidence. The only chemical samples available are a few rocks in the crust that have been carried up from the topmost layers of the mantle. To form an estimate of the composition of the earth as a whole and of the core and mantle one must rely on the chemistry of extraterrestrial bodies, including stars, the sun and meteorites. The presumption is that such bodies have a composition essentially similar to the earth's. The approach involves the formulation of physical and chemical models for the origin of the solar system and the earth.

It is generally held that the solar system formed about 4.6 billion years ago through the gravitational collapse of matter that previously had been dispersed in interstellar space. A large precursor of the sun, a protosun, was surrounded by a thin, disk-shaped nebula

of dust particles and gas. Local aggregation of particles and condensation of gas within the rotating nebula formed small objects that combined to produce planetary bodies. The material of meteorites was also formed during this period. Since the sun accounts for more than 99.6 percent of the mass of the solar system, its composition is effectively the same as that of the system as a whole. The abundances of the elements in the sun and other stars have been measured by spectroscopic methods, whereby each element is identified by its characteristic electromagnetic radiation.

What Meteorites Reveal

Meteorites now travel through the solar system in elliptical orbits that occasionally intersect the earth. There is evidence that they come from the asteroid belt: the swarm of small planetlike bodies orbiting at distances of 2.2 to 3.2 astronomical units from the sun. The meteorites vary widely in chemistry, mineralogy and structure, but for the purposes of this discussion it is sufficient to note the distinctions between two main groups: the iron meteorites and the stony meteorites. The iron meteorites consist essentially of iron-nickel

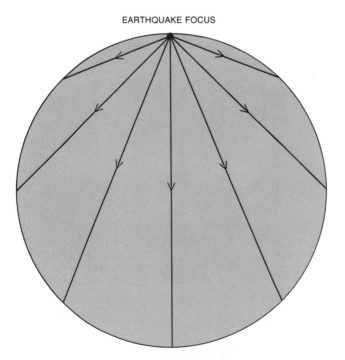

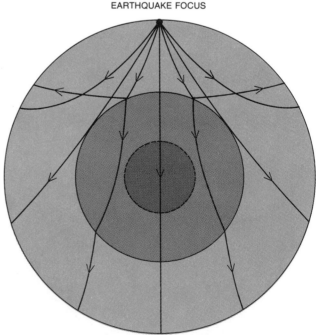

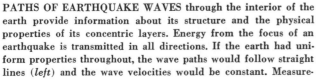

PATHS OF EARTHQUAKE WAVES through the interior of the earth provide information about its structure and the physical properties of its concentric layers. Energy from the focus of an earthquake is transmitted in all directions. If the earth had uniform properties throughout, the wave paths would follow straight lines (*left*) and the wave velocities would be constant. Measure-ments of the time of travel show, however, that the wave velocities, and therefore the physical properties, change abruptly at certain levels (*right*), thus revealing the concentric layers. *P* waves can be transmitted through both solids and liquids. *S* waves, which cannot pass through liquids, do not pass through the core of the earth, showing that at least the outer core is in the liquid state.

alloy, with the nickel content ranging from 4 to 20 percent; they also contain a small amount of iron sulfide.

Stony meteorites are composed mainly of silicate minerals, together with various proportions of metal alloy and iron sulfide. The relative abundances of nonvolatile elements such as magnesium, silicon, aluminum, calcium and iron are about the same in many types of stony meteorite as they are in the sun and other stars. It is therefore argued that these abundances provide a good basis for estimating the overall abundances of elements in the earth and other planets.

In complex models of the origin and evolution of the solar system the composition of the earth is derived by starting with a volatile-rich stony meteorite and formulating a series of processes and chemical changes that could account for the other types of meteorite and for the present structure of the earth. The calculations yield estimates of the composition of the core and the mantle.

A simpler approach, which gives a similar result, is to select a specific group of stony meteorites and to assume that the average composition of the silicate portion is equivalent to the composition of the earth's mantle. The composition of the earth's core is estimated from the iron sulfide and an appropriate portion of the iron-nickel alloy in the same meteorites. The iron meteorites serve as a check on this calculation.

Whatever procedure is followed, the estimates of the composition of the mantle agree in the following respects: (1) More than 90 percent by weight of the mantle is represented by oxides of silicon, magnesium and iron (SiO_2, MgO and FeO), and no other oxide exceeds 4 percent. (2) The oxides of aluminum (Al_2O_3), calcium (CaO) and sodium (Na_2O) total between 5 and 8 percent. (3) More than 98 percent of the mantle is represented by these six oxides, and no other oxide reaches a concentration of as much as .6 percent. The concentrations of other elements, which are present in trace amounts, are not defined. The oxides are combined in various minerals within the mantle rock.

Of all the rocks found in the crust, only peridotites correspond to the estimates of the composition of mantle rocks made by studying extraterrestrial bodies. It is therefore to such peridotites that geologists interested in the mantle turn for details, including the concentration of trace elements. The problem, of course, is to make sure that the specimen of peridotite one is examining originated

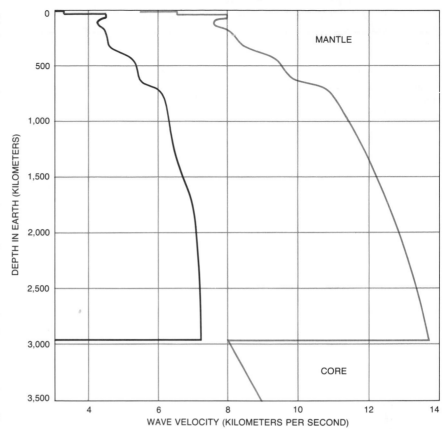

VELOCITY PROFILES of earthquake waves passing through the mantle are plotted for S waves (*black*) and P waves (*color*). The velocities are affected by the increase in pressure and temperature with depth, but the steps appearing at intervals down to a depth of about 1,000 kilometers correspond to changes in physical properties of the mantle. The decreased velocity between 100 and 250 kilometers is probably due to the presence of partly melted rock, which is an appropriate property for mobile asthenosphere of plate-tectonic model.

in the mantle rather than in the crust. For this reason particular interest attaches to the rounded boulders or nodules that are found in kimberlite "pipes" (cylindrical intrusions, with a diameter of a few hundred meters, that puncture the crust from the mantle in certain regions).

Kimberlites and Lavas

Kimberlites are famous as the rocks that bring diamonds to the surface of the earth. Diamond is a form of carbon that is stable only at very high pressures. One can therefore be confident that kimberlites originate in the depth interval from 150 to 300 kilometers below the surface—well within the upper mantle.

Evidence indicates that a kimberlite pipe originally rose rapidly through the crust as a fluidized system of solids, molten rock and gases, breaking through to the surface with a tremendous blast in a brief volcanic explosion. Fragments of rock ripped from the walls of the pipe and carried upward include nodules

of peridotite and subordinate nodules of eclogite, another mantle rock. They are rounded and polished by repeated impact with other gas-driven fragments in the explosive pipe.

Suites of nodules similar to those in kimberlites are also found in certain volcanic lavas. In general they are derived from shallower levels of the mantle than the nodules in kimberlites. Many estimates of the composition of the mantle have been based on the mineralogy and chemistry of nodule suites in kimberlites and lavas. Peridotite rocks found in other geological environments have also served for this purpose, although with them it is particularly important to take into account the nature of the geological association with other rocks and the effect of geological processes in order to be sure that the judgments made on the basis of the evidence in the rocks truly relate to the mantle rather than to events in the crust.

A third approach, in addition to the study of extraterrestrial bodies and mantle rocks, is to calculate a hypothetical peridotite having the chemistry appropriate to yield the volcanic lavas that are derived by partial melting of the mantle. The melting gives rise to the basaltic magma erupted abundantly onto the surface from volcanoes and leaves behind in the mantle a residual peridotite, which can be presumed to lack the more fusible elements that melted and were carried off in the magma. Assigning an appropriate chemistry to the residual peridotite, one arrives at the hypothetical composition of the upper mantle. Pyrolite (pyroxene-olivine rock) is the name given to one of these hypothetical peridotites.

The Heterogeneous Mantle

It was once assumed on the basis of physical models that the upper mantle was homogeneous in composition. As we have seen, however, the detailed geophysical studies of recent years indicate a layered structure. Indeed, the nodules that have been studied confirm the notion that the upper mantle is quite heterogeneous, both chemically and mineralogically. The specimens sample the mantle from the mantle-crust bound-

ary to the sources of kimberlite pipes, a depth interval that could reach 250 kilometers. One could hope to find among the nodules specimens of original mantle peridotite that has never been melted; specimens of depleted residual peridotite from which a molten fraction of the more fusible components has been removed; specimens of the molten fraction that failed to escape to the surface in magma, instead crystallizing at high pressure in the mantle to form eclogite; specimens intermediate between these types, and specimens involving the same mineral assemblage but formed by other processes too complex for consideration here.

Estimates of whole-mantle composition based on the examination of extraterrestrial bodies yield values similar to those obtained from the study of rocks originating in the upper mantle. Together the estimates support the idea that the chemistry of the mantle does not change much from top to bottom. In comparing estimates, however, one finds that the potassium content of the upper mantle according to the hypothetical pyrolite is considerably higher than the amount of potassium found in peridotite rocks de-

rived from the mantle, and both are much lower in potassium than the estimate derived from studying extraterrestrial bodies.

Large uncertainties also remain about the concentration and distribution of other trace elements and of volatile components such as water and carbon dioxide. These components are of fundamental significance for such factors as the generation of heat by radioactive decay (of such elements as uranium and thorium and the radioactive isotope of potassium, potassium 40), melting temperature (in the presence of small amounts of water at high pressure rocks begin to melt at temperatures lower than the melting point of dry rock) and the physical strength of the mantle (where the strength of rock would be significantly reduced by the presence of small amounts of molten material between mineral grains and by interstitial bubbles of gas).

Water and Carbon Dioxide

In certain nodules of peridotite one finds the minerals phlogopite and amphibole, which are hydrous, that is, incorporate water. This finding is accepted as evidence for the existence of water in at least some parts of the upper mantle. It is doubtful that the amount could exceed .1 percent by weight, and it is probable that the distribution of the water is not uniform.

Examination under the microscope has shown that the crystals of olivine and pyroxene in some peridotite nodules from kimberlites and lavas are crowded with tiny cavities up to five micrometers in diameter. Many of them are filled with dense, liquid carbon dioxide trapped at high pressure. This finding indicates the presence of carbon dioxide in some form in the upper mantle.

High-voltage electron microscopy has recently yielded remarkably detailed pictures of crystal defects: the discontinuities in crystal structure that occur within individual grains of minerals. The defects cannot be seen with light microscopes. In the electron micrographs minute bubbles of carbon dioxide are abundant along the discontinuities in the olivine and pyroxene of some peridotite nodules. This evidence suggests that the carbon dioxide was originally dissolved in the solid minerals and was exsolved and precipitated as gas bubbles because of elastic strain near the crystal defects.

The independent determinations of the physical properties and chemistry of

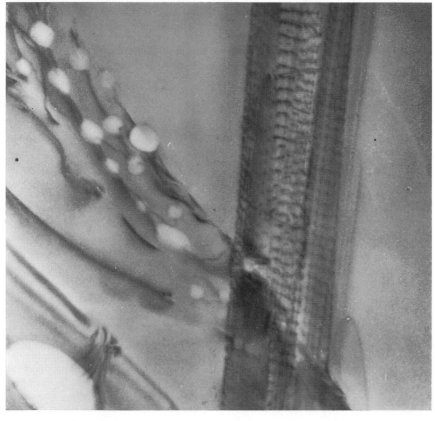

CARBON DIOXIDE BUBBLES appear in this electron micrograph of a specimen of peridotite. The enlargement is 20,000 diameters. The bubbles, which appear in discontinuities in the crystal structure of the rock, indicate the presence of carbon dioxide in the mantle. Micrograph was provided by Harry W. Green II of the University of California at Davis.

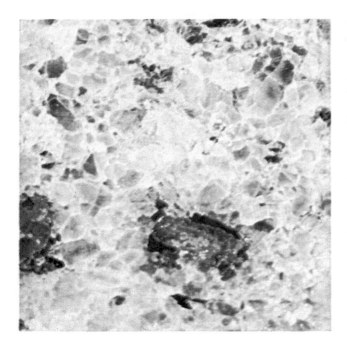 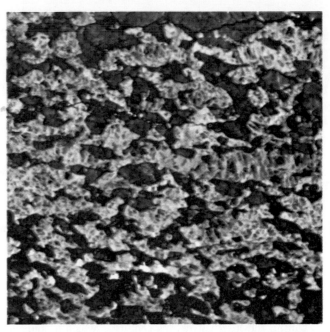

ROCKS FROM MANTLE are a mica-garnet peridotite (*left*) and an eclogite (*right*). The peridotite consists mainly of olivine (*yellowish green*) but also contains garnet (*red*) and orthopyroxene and clinopyroxene (*bright green*). A sample including mica, a mineral with water bound in its structure, is rare. This sample is from Tanzania. The eclogite, which is from a mine in South Africa, is composed of garnet (*red*) and a clinopyroxene (*green*). The minerals are arranged in layers. Mantle rocks are brought to the surface in kimberlite "pipes" and in certain volcanic lava. The photographs were provided by J. B. Dawson of the University of St. Andrews.

the mantle must be consistent with one another. In order to test their consistency one needs to know the physical properties of the estimated composition of the mantle through the range of pressure and temperature existing in the mantle. If one can determine the way the mineralogy varies as a function of pressure and temperature, one has a means of estimating the depth from which a rock sample came and the temperature at that depth when the rock was formed or reached equilibrium with its surroundings.

Pressure and Temperature

The structure of silicate minerals is dominated by the packing of oxygen atoms. The other atoms, which are much smaller, occupy spaces between the oxygens. At low pressure each silicon atom is surrounded by four oxygens whose centers form a tetrahedron; the silicon is said to be in a fourfold coordination. At much higher pressure the oxygen atoms are squeezed closer together. They adjust into a densely packed arrangement with silicon atoms in a sixfold coordination. This readjustment of mineral structure is a phase transition. The steplike changes in value of the physical properties in the mantle are caused by successive phase transitions.

Starting with a material of fixed com-

position, such as the peridotite thought to exist in the mantle, the phase transitions depend on pressure and temperature. A diagram of the transitions for peridotite has been determined by experiments in laboratory apparatus that carried the pressure up to 200 kilobars (200,000 atmospheres, equivalent to the pressure 600 kilometers below the surface of the earth). The diagram has been extended to higher pressures by indirect methods.

It is known from the nodules in kimberlites and lavas that peridotite in the mantle can crystallize in at least three mineralogical assemblages: plagioclase peridotite, spinel peridotite and garnet peridotite. The high-pressure experiments demonstrate that the assemblages are related through phase transitions. With increasing pressure plagioclase peridotite is transformed first into spinel peridotite and then into garnet peridotite [*see illustration on next two pages*].

Experimental studies show that at still higher pressure garnet peridotite undergoes a phase transition involving an increase in density of almost 10 percent; the dominant olivine of the upper mantle is transformed into a spinel-like material, and the aluminous pyroxene is transformed into a garnet structure that combines in solid solution with the garnet already present. At pressures near 200 kilobars the minerals are further

compressed into structures with all the silicon atoms in the sixfold coordination, giving rise to minerals that are unknown at the surface of the earth. This compression results in another increase in density of about 10 percent. The actual pressure at which a phase transition occurs increases with higher temperatures.

At any fixed depth an increase in temperature eventually brings a rock to a point where it begins to melt. This temperature increases with pressure, as shown by the boundary labeled solidus in the phase diagram [*next two pages*]. A rock composed of several minerals melts progressively through an interval of temperature in which solid crystals coexist with liquid. Complete melting is marked by the boundary labeled liquidus.

The effect of temperature can be studied by means of a geotherm, which is a line giving the temperature at each depth in the earth. If the line is drawn on the phase diagram for peridotite, each point on the line occupies one of the phase fields and thus also defines the mineral assemblage for the peridotite at each depth. A cross section through a hypothetical mantle composed of peridotite is constructed by following a geotherm through the phase diagram. Each layer consists of a particular mineral assemblage.

The boundaries between layers of the mantle are presumably at depths where the geotherm crosses phase boundaries. It turns out that these boundaries correspond closely to the depths where the velocity of seismic waves changes. This finding is regarded as being good evidence that the upper mantle does have a composition close to that of the hypothetical peridotite and that the layered structure of the upper mantle is caused by transitions of phase rather than by changes in composition.

The decrease in the velocity of earthquake waves in the low-velocity zone can be explained by the presence of water or carbon dioxide in the upper mantle. Either one could cause a trace of melting in the peridotite of the upper mantle. The result is a change in the physical properties of the rock. If no water were present, a similar effect could perhaps be produced by intergranular carbon dioxide.

During the first half of this century it was widely believed among geophysicists that convection, the upward flow of hotter material and the downward flow of cooler material, could not occur in the solid, rigid mantle. That was one of the reasons the theory of continental drift failed for so long to gain many adherents. Recently, however, a number of models have proposed convection in the mantle as the driving mechanism for the migration of lithospheric plates. Details of the movements in the mantle and the scale of convection remain uncertain, but there is little doubt that the rates are exceedingly slow—so slow that the mantle is effectively motionless within the normal human framework of time.

The lithospheric plates and the continents that ride on them drift at the rate of a few centimeters per year. Suppose the driving material in the mantle moves at a rate of five centimeters per year, which is equivalent to about .005 milli-

meter per hour. The tip of the hour hand on a typical household clock moves five centimeters per hour, and the movement is not directly apparent to the human eye. Yet the speed is 10,000 times faster than the proposed movement in the mantle. Even so, a movement of five centimeters per year adds up significantly over geologic time; a parcel of rock could move from the bottom of the mantle to the top in 58 million years, which is only a small fraction of the earth's age of 4.6 billion years.

Movements of the Mantle

How is it possible that solid rock can flow, even so slowly? When a blacksmith picks up a bar of cold steel, he is unable to bend it, but when he heats it to a red glow, he can bend it easily, even though it is still a solid bar. Similarly the rocks of the mantle, which are at

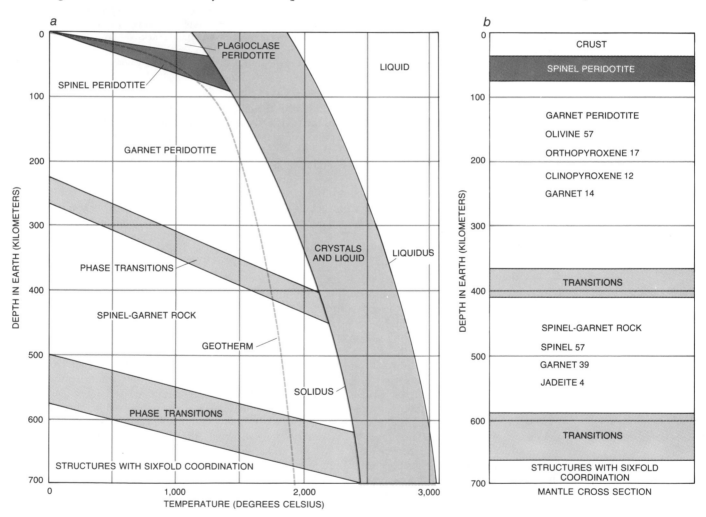

PHASE DIAGRAM for peridotite is stated in terms of pressure, or depth, and temperature. At a fixed depth an increase in temperature eventually brings a rock to a point where it begins to melt. The beginning and end of melting are bounded by the solidus and liquidus curves. Reading downward in the two parts of the diagram at left, following the geotherm (a), one sees that the upper mantle has a layered structure owing to successive phase transitions from spinel peridotite in the uppermost region to a rock composed of minerals with elements in a sixfold coordination, that is, having closely packed groups of oxygen atoms enclosing atoms of magnesium, iron and silicon, at a depth of about 600 kilometers. The numerals in the cross-section diagram (b) represent the percentages

high temperatures, can be deformed even though they are still in the solid state.

Olivine, pyroxene and peridotite have been subjected to strains at high pressures and temperatures in laboratory experiments. They deform. The deformed products have recently been studied in the high-voltage electron microscope in attempts to establish the mechanisms of plastic flow. A suggestion emerging from this work is that deformation occurs first in individual crystals, which then recrystallize to form a mosaic of new grains. The mechanism of flow varies as a function of pressure and temperature.

Schemes have been proposed for large convection cells extending through the entire thickness of the mantle [*see illustration on page 96*]. An alternative model, based on the argument that the absorption of heat in phase transitions would reduce the driving force, confines

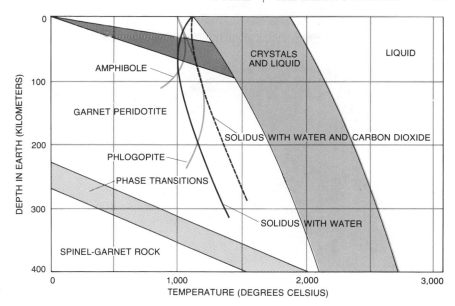

LOW-VELOCITY LAYER in the uppermost mantle appears to be explained by the presence of water and carbon dioxide, which lower the solidus curve, so that rock melts at lower temperature and pressure. The hydrous, or water-bearing, minerals amphibole and phlogopite are stable through a limited temperature range (*color*). Comparing this diagram with the one for dry mantle rock, one sees that the geotherm curve passes from a phase of solid peridotite to a phase of partly melted peridotite at a depth of about 100 kilometers, which corresponds closely with the top of the low-velocity zone observed in the uppermost mantle.

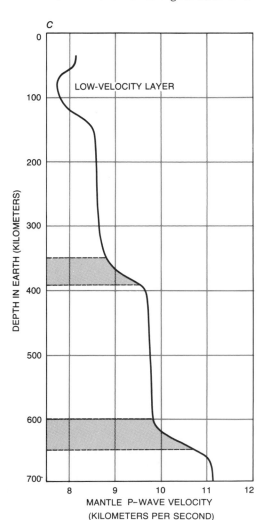

of minerals in the rocks. The profile of *P*-wave velocity (*c*) shows that the depths of actual changes in the mantle indicated by earthquake waves coincide closely with the depths of phase transitions as inferred from the phase diagram and the geotherm.

convection to the upper mantle, above the olivine-spinel transition. A third model limits convection to the asthenosphere, which as a layer of the mantle lies at depths of from 100 to 300 kilometers. Subduction zones, however, where lithospheric material returns to the mantle and thus is involved in convective forces, appear to run as deep as 700 kilometers. Another model for the plate-driving mechanism involves thermal "plumes" in the mantle. According to this argument, all upward movement of mantle material is confined to about 20 plumes, each plume a few hundred kilometers in diameter, rising from the core-mantle boundary. The return flow is accomplished by a slow downward movement of the rest of the mantle.

Where a plume reaches the lithosphere the flow becomes horizontal, spreading radially in all directions. A plume creates a hot spot with volcanic activity at the surface, and it may cause an upward doming of the lithosphere. In such ways the plumes would cause the movement of the lithospheric plates.

The hot-spot hypothesis is currently a hot idea. Many geologists are exploring its implications for various phenomena, including the features of volcanic island chains such as Hawaii. The hypothesis is also under challenge, however, by earth scientists who express doubt that a plume could retain its coherence as it rises through 2,800 kilometers of mantle.

The vertical movement of mantle ma-

terial in any kind of convective system causes changes in the temperature distribution; at a given depth the temperature is increased where hotter material is rising and decreased where cooler material is sinking. The movement changes the shape of the geotherm curve from time to time and from place to place as the convection proceeds. The compositions of the minerals in mantle peridotite vary as a function of pressure and temperature, as laboratory experiments dealing with the phase diagrams of peridotite and its constituent minerals have established.

A parcel of mantle peridotite is subjected to changes in temperature and pressure as a result of convective motions in the mantle, and its mineralogy will adjust by recrystallization as it strives to attain equilibrium in the changing environment. The movements are slow enough so that an equilibrium mineralogy is normally attained. If the rock is suddenly transported to the surface, however, as in a kimberlite eruption, the time for readjustment is not enough and the sample reaches the surface with the mineralogical signature corresponding to its position and temperature where it last reached equilibrium in the mantle.

The compositions of coexisting minerals in peridotite at any particular pressure and temperature have been measured directly in the laboratory experiments. Using these data for the purpose

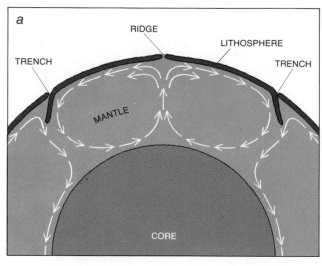

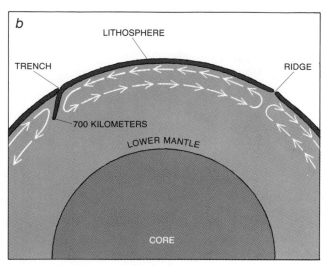

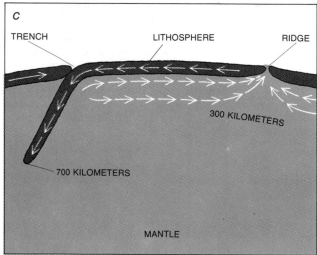

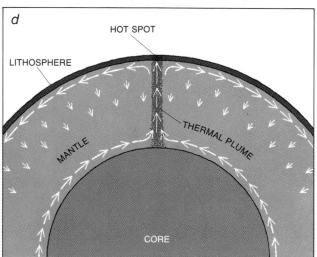

MODELS OF CONVECTION have been proposed to explain how activity in the mantle drives the lithospheric plates. In convection warmer material moves upward and colder material moves downward. One model (a) holds that convection cells extend through the entire mantle. According to a second model (b), they are confined to depths above the phase transition from spinel to olivine. A third model (c) confines movements of the mantle to the asthenosphere. In the thermal-plume model (d) all upward movement is confined to a few thermal plumes, and the downward flow is accomplished by slow movements of the remainder of the mantle.

of calibration, it is now possible to take samples of mantle peridotite, such as the nodules from kimberlites, to measure the composition of the minerals in each sample and thus to estimate the pressure (or depth) and temperature at their source in the mantle.

Ancient Geotherm

The application of these methods to nodules from kimberlites has produced intriguing results during the past year or two. A suite of nodules collected from a single kimberlite pipe gives a series of points, each specified for a given nodule by the estimated pressure and temperature of equilibrium before eruption. The locus of these points on a diagram of depth (pressure) and temperature corresponds to the geotherm that existed at

the time of eruption of the kimberlite. In other words, each kimberlite nodule contains stored within its mineralogy the record of its equilibrium pressure and temperature in the mantle before it was abruptly carried upward. The experimental mineralogists engaged in this work have found that the results from kimberlite pipes in South Africa give fossil geotherms with normal gradients down to depths of 150 kilometers or so but with steeper gradients at deeper levels; apparently the temperature was higher than normal at those depths for some interval of time before the kimberlite pipes erupted nearly 100 million years ago.

These results are of great interest to geophysicists trying to establish the thermal history of the earth, the dynamics of the mantle and the driving forces

of plate tectonics. One interpretation of the data is that the inflection point in a fossil geotherm corresponds to the top of the asthenosphere at a time about 120 million years ago when the African lithospheric plate began to move rapidly as the Atlantic Ocean opened up. According to this interpretation, frictional heating in the asthenosphere as a result of movement shifted the geotherm to higher temperatures, producing an inflection point at the lithosphere-asthenosphere boundary, whereas the conduction rate of heat through rocks is so slow that the original geotherm in the lithosphere remained effectively unchanged. Another interpretation is proposed by geophysicists who conclude that friction in the asthenosphere is not capable of producing such a large thermal effect. They argue that the steepened portion of the fossil geotherm below the inflection point could be caused by the upward convection of a local thermal plume, which initiated the eruption of the kimberlite.

This example illustrates the way plate tectonics, with its focus on the mantle, has brought together into the same symposium rooms at scientific meetings research workers in areas that once were considered quite distinct from one another. Field geologists, mineralogists, geophysicists and experimental chemists and physicists have together discovered in the minerals of kimberlite nodules information that is grist for the mills of the theoreticians who are trying to work out how the mantle moves now and has moved through the 4.6 billion years of the earth's existence.

A view of the entire earth from a spacecraft shows a large, spinning globe as smooth as a billiard ball. The deepest hole drilled in the surface reaches a depth of only nine kilometers; it is a mere pinprick, penetrating less than .15 percent of the earth's radius. It is therefore quite remarkable that so much is known about the inaccessible mantle.

Uncertainty of Models

Nonetheless, the amount of information now in hand is insufficient for a full understanding of the dynamic behavior of the mantle, which is the key to many geophysical and geological phenomena. The extent of the uncertainty about what is really happening in the mantle can be illustrated by comparing two hypotheses about the Hawaiian volcanic island chain. Each island formed as a result of eruption above a melting region fixed in the asthenosphere. Then the moving lithospheric plate carried the island away, creating over a long period of time the island chain.

According to one interpretation of these events, the melting is localized in a hot spot above a thermal plume. According to another interpretation, the melting is localized by friction in rocks flowing from the asthenosphere into a column moving downward through the mantle. The column is said to form a gravitational anchor, maintaining the region of downward flow in a position more or less directly above it.

These diametrically opposed hypotheses are put forward by respected earth scientists. Is Hawaii to be explained by a rising thermal plume or by a sinking gravitational anchor? I am confident that before too many years have passed the accumulation of additional evidence and the refinement of hypotheses will have placed much closer constraints on the picture that can be drawn of the structure and dynamics of the mantle.

8

Kimberlite Pipes

by Keith G. Cox
April 1978

These remarkable fossil volcanoes rise from a great depth. They are the ultimate source of diamonds and also of rocks that may be specimens of materials from the earth's mantle

Living on the surface of the earth, geologists have little direct knowledge of the planet's interior. Of the three broad layers that make up the earth's structure—the crust, the mantle and the core—only the crust is accessible, and even in its thickest regions the crust represents only about 1 percent of the earth's radius. Certain physical characteristics of the deeper layers, such as their average density and the speed with which they transmit earthquake waves, can be deduced from the surface. For studies of chemical composition, however, there is no adequate substitute for a specimen of mantle material.

An extraordinary source of such specimens is the rare rock type called kimberlite. Kimberlite formations generally take the form of small vertical shafts, called pipes, which are demonstrably of volcanic origin. The pipes have been studied extensively, in large part because they are of economic importance: they are the ultimate source of natural diamonds. For the geologist, however, kimberlite pipes supply gems of a different kind: rocks brought up from a great depth. Some of these rocks may be samples of material characteristic of that found in the upper portions of the earth's mantle.

Until about 100 years ago the only known deposits of diamonds were in river gravels. In 1870, however, alluvial diamond deposits in southern Africa were traced to their source, the kimberlite pipes of Jagersfontein and Dutoitspan. The pipes were near a town that is now the South African city of Kimberley, and the characteristic rock type in which the diamonds are found was named for the town.

Several other pipes have since been discovered at Kimberley, and isolated pipes and small groups of pipes are scattered in other parts of southern Africa. There is a group of 17 pipes in Lesotho, the small country surrounded entirely by South Africa, and others are known in Botswana, Namibia, Angola and, to the north, in Tanzania. Elsewhere in the world the only comparable concentra-tion of kimberlite deposits is in the Yakutsk Republic in Siberia; the kimberlites were discovered there only in 1954. In North America there is a concentration of pipes along the border between Colorado and Wyoming, and others are known in Montana and in the Canadian Arctic. Few of the American kimberlite pipes are large or economically important, although a solitary pipe at Murfreesboro, Ark., was briefly worked as a diamond mine.

Compared with the commoner remnants of volcanic activity on the earth's surface, kimberlite pipes are quite small features. The largest have diameters at the surface of less than two kilometers, and many pipes of economic importance are only a few hundred meters in diameter. The pipes generally have the form of a cylinder or a narrow cone that tapers slightly with increasing depth. In the vicinity of the pipes kimberlite can also be found in associated formations called dykes, which are vertical slabs formed by the intrusion of molten material into fissures in the surrounding rocks. The pipes probably erupted at the surface when they were formed and were then marked by an open crater and a small cone of ejected material. In almost all cases, however, subsequent erosion has removed the surface features and the uppermost strata of both the kimberlite and the surrounding rocks. The pipes now available for study are exposed at deeper erosion levels.

Diamonds are released from the kimberlite by erosion, and they generally settle in stream beds. Subsequent geological changes may bury and consolidate these alluvial deposits, but the diamonds, being extremely durable, remain unaltered. Most of the known kimberlite pipes were emplaced in the Cretaceous period, some 70 million to 130 million years ago. Diamonds are found in alluvial deposits of several geological ages, however, indicating that there were also pipes in earlier periods. For example, there are extensive alluvial diamond fields in Brazil that are not associated with any known kimberlites. Presumably the older pipes are now hidden by younger, overlying formations. Curiously, one of the largest pipes in South Africa, the Premier, is dated at more than 1,150 million years and hence is much older than the characteristic Cretaceous pipes of the region.

Kimberlite is a highly variable rock type. Most kimberlite exposed at the surface, called "yellow ground" by miners and prospectors, is severely weathered. At deeper levels there is a material that is better preserved called "blue ground," but only in recent years have samples of the native kimberlite become readily available. Fresh kimberlite is a hard, dark gray or blue rock whose structure gives unmistakable evidence of an igneous origin. The kimberlite was extruded into its present position as a molten liquid; it was then cooled by contact with the volcanic conduit and finally solidified.

The major constituents of kimberlite are silicates, that is, compounds of silicon and oxygen with metal ions. In general, minerals cannot be defined as simple chemical compounds because their composition is not determined by a fixed ratio of atoms. Often two or more compounds are present and are said to be in solid solution with one another. As in a liquid solution, the component substances can be mixed in any ratio over a wide range. One important constituent of kimberlite is the mineral called olivine, which is a solid solution of magnesium silicate (Mg_2SiO_4) and iron silicate (Fe_2SiO_4). Another silicate present is phlogopite, a kind of mica rich in potassium and magnesium, and there are also various silicate minerals that are classified as serpentines. The serpentines are formed by the hydration of olivine, or in other words by chemically adding water to it. Kimberlite also contains the mineral calcite, which is not a silicate but consists of more or less pure calcium carbonate ($CaCO_3$).

Of the materials found in kimberlite pipes kimberlite itself may be less interesting than some of the foreign bodies that appear as inclusions within the kimberlite matrix. Among these inclusions, of course, are diamonds, and it is to their presence that we owe much of our knowledge of these remarkable volcanoes. Without the economic incentive for exploration and mining that diamonds provide, few samples of kimberlitic rocks would be available for study.

Another kind of inclusion in kimberlite, and one that is far commoner than diamond, consists of rocks torn loose from the walls of the volcanic pipe. As the molten kimberlite approaches the surface and the pressure on it is thereby reduced, dissolved gases (mainly water vapor and carbon dioxide) are forced out of solution. As a result the volcano erupts explosively. The sides of the conduit are abraded, an effect that is intensified near the surface, so that the pipe grows wider near the mouth. Some of the wall rock is probably ejected during the eruption, but much of it is shattered, ground up and incorporated into the kimberlite. These inclusions are called xenoliths (from the Greek for foreign rocks).

Because the xenoliths derive from rock strata that can be observed in the terrain surrounding the pipe, their depth of origin can be determined. Many of them fell into the pipe and are found hundreds of meters below the equivalent rocks in the pipe walls. Xenoliths found in blind offshoots of the pipe (that is, in tubes that did not reach the surface) indicate that some fragments fell down and then were carried upward again.

Perhaps the greatest scientific interest in kimberlites derives from a third kind of inclusion: the rocks called ultramafic nodules. Like diamonds, the ultramafic nodules are rare. Also like diamonds, they are thought to come up from a great depth, perhaps as much as 250 kil-

KIMBERLITE PIPE in South Africa has been extensively excavated by open-cast mining for diamonds. The kimberlite deposit itself has been removed to a depth of a few hundred meters, so that the form of the original deposit is revealed by the size and shape of the cavity. Compared with commoner types of volcano, kimberlite pipes are **quite small geological features; they generally range from a few hundred meters to two kilometers in diameter. In the vertical dimension, however, these narrow shafts extend through the earth's crust and into the upper mantle and they may have carried rocks from that region to the surface. Pipe shown is the Premier, near Johannesburg.**

ometers below the surface. They have a characteristic rounded form, like beach stones, caused by abrasion in the pipe.

An important question is why ultramafic nodules appear in kimberlite pipes when they are almost never seen in other volcanoes, some of which may also originate deep within the mantle. One hypothesis is that the kimberlites rise through the earth much faster than the magmas of other volcanoes. If the fluid rock behaves like an ordinary liquid, such as water, then a rapid ascent would be required to transport the largest of the nodules. Such liquids are said to have zero shear strength: one part of the fluid can move freely past another part. Hence a solid body immersed in the fluid can be carried upward only if the force of drag created by the moving fluid exceeds the weight of the nodule. Calculations based on the weight of the largest nodules suggest that they may have been carried to the surface in a period of hours or at most a few days.

An alternative to the rapid-ascent hypothesis has recently emerged. R. S. J. Sparks, H. Pinkerton and R. MacDonald of the University of Manchester have proposed that under some circumstances molten silicates may not behave like ordinary fluids; they may have a shear strength greater than zero. The ultramafic nodules might then be transported not by the force of drag but as fixed inclusions in the fluid matrix. A peculiar property of such fluids is that they retain their shear strength only when they are flowing at low speed, and so the alternative to a rapid ascent is an exceptionally slow one.

Physical conditions inside the earth and certain physical properties of materials found there can be deduced from surface measurements. For example, since the mass and the angular momentum of the earth are known, the distribution of density can be calculated. From that information the internal pressure can be determined as a function of depth. It is assumed that the pressure is hydrostatic, or in other words that it is exerted equally in all directions by material that is not itself compressible. It can then be shown that the pressure increases by about one kilobar, or 1,000 times atmospheric pressure at sea level, for every three kilometers of depth. Internal temperatures can be estimated by measuring the flow of heat near the surface. Temperature increases with depth everywhere, but the increase is slower under continents than it is under the sea floor.

The most revealing view of the earth's interior comes from the observation of earthquake waves propagated over long distances. The refraction of these waves at various depths reveals sudden changes in their speed. Those changes may correspond to discontinuities in other properties as well.

The division of the earth into crust, mantle and core is based on such seismic observations. The crust is some 10 kilometers thick under the oceans, and it ranges from 35 to 70 kilometers in thickness under the continents. The mantle extends from the base of the crust to a depth of about 2,900 kilometers, and the core extends to the center of the earth at a depth of 6,370 kilometers. The mantle can be further divided into upper and lower layers at a depth of about 700 kilometers.

Seismic observations have revealed

ULTRAMAFIC NODULES are rare inclusions in kimberlite pipes that seem to have been transported in the solid state from the upper mantle. Several small, broken nodules are embedded in the kimberlite matrix of the rock at left; the rounded boulder in the center is a single large nodule. The smooth, rounded forms characteristic of the nodules are produced by abrasion during transport. The term "ultramafic" describes rocks made up mainly of magnesium and iron silicates. The commonest class of nodules, which includes the ones shown here, is composed of the rock type called peridotite, which is thought to be a major constituent of the mantle.

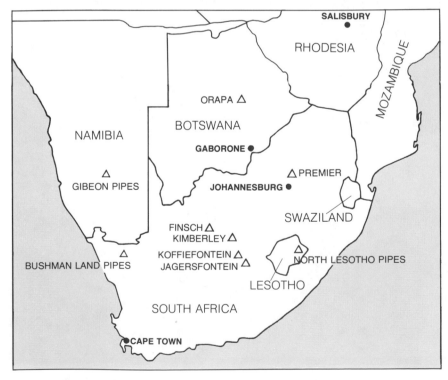

ABUNDANCE OF KIMBERLITE PIPES is greater in southern Africa than in any other region of the world. At most of the locations marked there is not just one pipe but a cluster of them; the North Lesotho group, for example, includes 17 pipes. The only concentration of pipes comparable to the one in southern Africa is in Siberia. Perhaps significantly, both southern Africa and Siberia were sites of extensive volcanic activity before the pipes were formed.

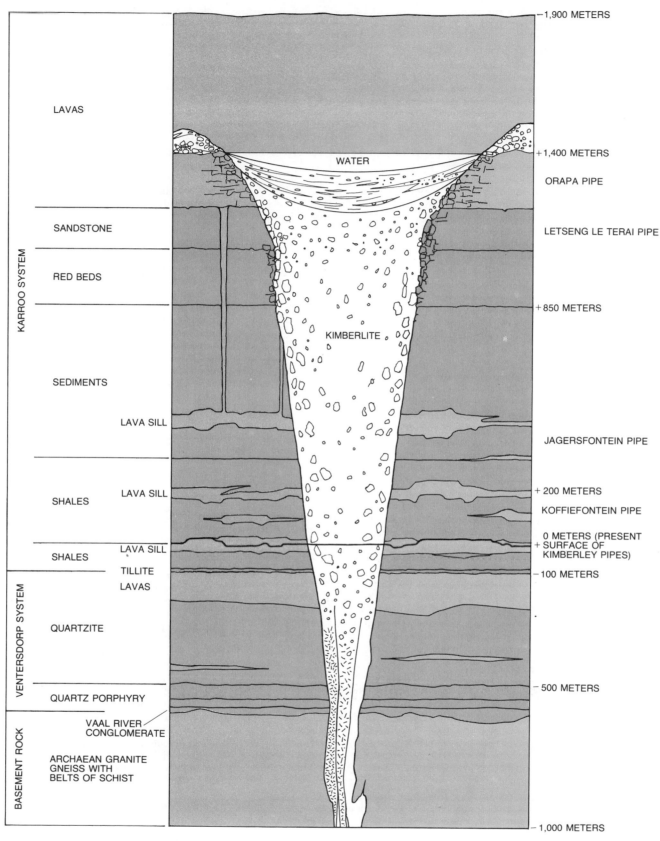

MODEL OF A KIMBERLITE PIPE is based on the structures of several pipes exposed at various erosion levels. The pipes erupted in the Cretaceous period, some 70 to 130 million years ago, when the strata that are exposed at Kimberley today were 1,400 meters below the surface. The geological structure of the region consists of an alternating sequence of igneous rocks (*color*) and sedimentary rocks (*gray*); both the igneous and sedimentary layers include a variety of rock types. The eruption of the kimberlite probably produced a shallow crater, surrounded by a low cone of ejected material. Abrasion of the wall rocks by the ascending kimberlites, which became more intense near the surface, gave the pipe the form of a narrow conc. Both fragments of the wall rock and the much rarer ultramafic nodules are incorporated into the kimberlite. Several African pipes (*noted at right*) are exposed at higher levels than those near Kimberley, the South African town that is the namesake of the rock type. The model was devised by J. B. Hawthorne of De Beers Consolidated Mines Ltd.

another discontinuity within the upper mantle, which serves as the basis for an independent and complementary interpretation of the earth's structure. The material above this region, including all of the crust and a part of the upper mantle, is the lithosphere. It consists of rigid plates, several of which are in relative motion. The zone below is the asthenosphere; it is distinguished by lower viscosity, and it may be subject to partial melting. The plates of the lithosphere ride on and slide across the comparatively fluid asthenosphere.

Some of the physical discontinuities deduced from seismic observations surely correspond to boundaries of chemical composition. In the continental crust the dominant constituents are granites and related rock types. They are rich in silicon, aluminum, sodium and potassium, and are comparatively poor in magnesium and iron. The oceanic crust is made up largely of basalt, which has more magnesium and iron than granite, but is poorer in sodium and potassium. Slices of oceanic crust are occasionally thrust upward in collisions between lithospheric plates, and even more occasionally they carry with them a layer of material from the topmost few kilometers of the mantle. The mantle rocks exposed by such upthrust plates are predominantly of the type called peridotite. The peridotites have a composition that is close to being the inverse of the granites in the continental crust: they are very rich in magnesium but are slightly depleted of silicon and are considerably depleted of aluminum, sodium and potassium.

The best available information on the composition of the earth as a whole comes not from terrestrial rocks but from meteorites. The most primitive meteorites, the chondrites, have an unusual dropletlike texture. They are thought to be unaltered specimens of the material from which the solar system condensed. The chondrites consist of a silicate rock similar to peridotite in overall composition, together with metallic iron that contains a substantial amount of nickel in solution. If the average compositions of the chondrites and the earth are similar, then the mantle must be mainly peridotite and the core must be iron and nickel. (The crust makes up such a small part of the earth's bulk that it can be neglected.)

The ultramafic nodules of kimberlite pipes provide important evidence corroborating this hypothesis. Unlike meteorites they are actual samples of terrestrial material, and hence their interpretation does not depend on the adoption of a particular model for the evolution of the earth. Unlike the mantle material associated with upthrust plates they are from deep within the mantle and not from the boundary layer where the mantle meets the crust.

"Mafic" is a term applied to minerals that contain large amounts of the silicates of magnesium and iron; in ultramafic rocks these minerals are the dominant constituents. Most of the ultramafic nodules found in kimberlite are made up of peridotite; peridotite in turn is composed mainly of olivine and another silicate mineral, pyroxene.

The olivine in peridotite is rich in magnesium, so that in its composition Mg_2SiO_4 dominates over Fe_2SiO_4. The nodules contain two kinds of pyroxene, which can be distinguished not only by their composition but also by the symmetry systems of their crystals. One is an orthopyroxene and forms orthorhombic crystals; the other is a clinopyroxene and forms monoclinic crystals. The orthopyroxene, which is the most abundant mineral after olivine, consists of a magnesium silicate ($MgSiO_3$) with some iron silicate ($FeSiO_3$) in solid solution. The clinopyroxene is chrome diop-

STRUCTURE OF THE EARTH has been determined largely by seismic observations. The boundary between the crust and the mantle, which is at a depth of 10 kilometers under the oceans and from 35 to 70 kilometers under the continents, is marked by a discontinuity in the speed of earthquake waves. It probably also represents a boundary between regions of different chemical composition. The oceanic crust is mainly basalt and the continental crust mainly granite, whereas the mantle is apparently peridotite. Only the upper 1,000 kilometers of the earth's structure is shown; the mantle continues to a depth of 2,900 kilometers and the core extends from that level to the center of the earth at 6,370 kilometers. A seismic discontinuity at a depth of about 700 kilometers divides the mantle into upper and lower regions, but it probably represents only a change in crystal structure, not in chemical composition. A zone of low earthquake-wave velocities is the basis for another system of layers (left). Material above this zone is the rigid lithosphere, which slides over the more plastic asthenosphere. Ultramafic nodules in kimberlite are thought to have come from the mantle at depths of from 100 to 300 kilometers.

side, a bright green mineral that contains substantial amounts of calcium. Its elemental composition is approximately $CaMgSi_2O_6$, but it also contains some iron and chromium. Many peridotites also have small quantities of pyrope, a red-to-violet variety of garnet. The approximate composition of pyrope is $Mg_3Al_2Si_3O_{12}$, but it too has some iron in solution. Olivine and orthopyroxene are invariably the major constituents of peridotitic nodules. Chrome diopside and pyrope are usually present, although only in small quantities, but in some nodules they are absent entirely.

Although the peridotitic nodules are by far the most abundant type in most kimberlite pipes, other kinds of nodule are known. One of the most interesting is eclogite, which is composed largely of garnet and a calcium-rich pyroxene called omphacite. Eclogites have several intriguing properties, not the least of which is that they occasionally contain diamonds. (Most diamonds, however, are found not in eclogites but as single crystals within the kimberlite matrix.)

The predominance of peridotite in the ultramafic nodules lends strong support to the hypothesis that peridotite is the major constituent of the mantle. As I shall show below, the ultramafic nodules seem to have been brought up from a depth of 100 kilometers or more. Since there are no discontinuities in the physical properties of the mantle down to 700 kilometers, it seems reasonable to conclude that the entire upper mantle consists mainly of peridotite. In fact, it is likely that the lower mantle is also peridotite. The refractive zone at 700 kilometers probably marks not a change in composition but a mineralogical transition where high pressure alters the crystal structure of the peridotitic materials.

The depth at which an igneous rock was formed can sometimes be estimated from knowledge of the minerals it contains. The conditions required to create the mineral, and particularly the temperature and pressure, can be determined in the laboratory. Geophysical measurements then provide an estimate of where in the earth those conditions prevail. In kimberlites diamond provides an upper limit to the depth of formation and various forms of silica give a lower limit.

Diamond is a crystalline form of carbon that can be created only at high temperature and pressure. Under milder conditions the favored crystalline form of carbon is graphite. The set of all combinations of temperature and pressure where the two forms are in equilibrium is called the diamond-graphite inversion curve. A specimen of carbon at any point on this curve could assume either crystalline form; increasing the pressure or decreasing the temperature would favor the creation of diamond.

	NODULE COMPONENT	APPROXIMATE COMPOSITION	CRYSTAL SYSTEM
PERIDOTITE / PYROXENE	OLIVINE	Mg_2SiO_4 WITH Fe_2SiO_4	ORTHORHOMBIC
	ORTHOPYROXENE	$MgSiO_3$ WITH $FeSiO_3$	ORTHORHOMBIC
	CLINOPYROXENE (CHROME DIOPSIDE)	$CaMgSi_2O_6$ WITH Fe AND Cr	MONOCLINIC
ECLOGITE	GARNET (PYROPE)	$Mg_3Al_2Si_3O_{12}$ WITH Fe	CUBIC
	CLINOPYROXENE (OMPHACITE)	$CaMgSi_2O_6$ WITH Fe, Na AND Al	MONOCLINIC

MINERAL CONSTITUENTS of ultramafic nodules are mainly silicates, or compounds of silicon and oxygen with metal ions. In the commoner peridotite nodules the major constituents are olivine and orthopyroxene; most peridotites also contain small quantities of clinopyroxene and garnet, but those components are sometimes lacking. Another nodule type, eclogite, is made up of garnet and a clinopyroxene called omphacite. A distinguishing feature of nodule minerals is that they are much richer in magnesium than rocks typical of the earth's crust.

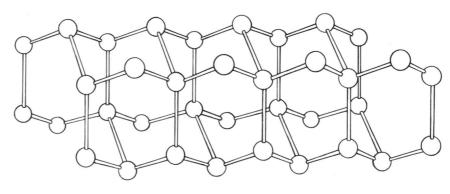

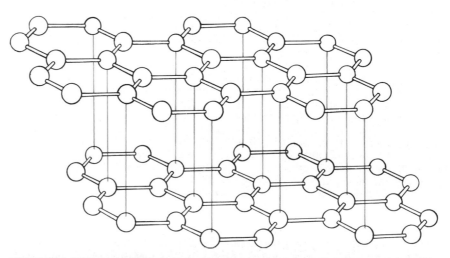

DIAMOND AND GRAPHITE are both crystalline forms of carbon, but they are stable at different combinations of pressure and temperature. Graphite, the low-pressure form, consists of stacked planes of hexagonal rings. Diamond, which can be formed only at high pressure, has a more symmetrical structure in which each carbon atom is surrounded by four others. The presence of diamond in kimberlite gives a clue to the conditions under which the rock was formed.

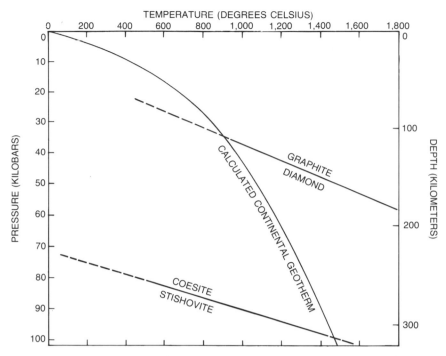

TEMPERATURE (DEGREES CELSIUS)

PRESSURE (KILOBARS)

DEPTH (KILOMETERS)

CALCULATED CONTINENTAL GEOTHERM

GRAPHITE
DIAMOND

COESITE
STISHOVITE

UPPER AND LOWER BOUNDS on depth of kimberlite formation are provided by forms of carbon and silica found in kimberlite pipes. The presence of diamond indicates that kimberlite forms at a depth below the diamond-graphite inversion curve, which marks the zone where these phases are in equilibrium. Similarly, kimberlite contains silica in the form of coesite but not as stishovite, and so it probably comes from a region above the stishovite-coesite inversion curve. Temperatures within the earth can be calculated from measurements of heat flow at the surface; the resulting curve for regions below continents is called the continental geotherm. Kimberlite and the nodules it contains are likely to have formed along the geotherm between those points where it intersects the diamond-graphite and stishovite-coesite inversion curves.

Temperature and pressure within the earth are highly correlated. The graph of temperature as a function of pressure or depth is called the geotherm; since kimberlites are known only on land, the curve of interest is the continental geotherm. The intersection of the diamond-graphite inversion curve and the continental geotherm gives the minimum likely depth for the formation of diamonds. The curves have been found to meet at a pressure of about 35 kilobars and a temperature of roughly 800 degrees Celsius. That pressure corresponds to a depth of about 105 kilometers. At all shallower levels the pressure is too low or the temperature is too high for diamond to form. The presence of diamond in kimberlite therefore indicates that the mineral formed at a depth greater than about 100 kilometers. That depth is only a minimum, however; since diamond is stable at all pressures below the inversion curve it could be formed at any greater depth.

A maximum depth for kimberlites can be calculated from the inversion curve for two crystalline forms of another substance: silica. At low pressure silica is represented by the familiar mineral quartz. As the pressure is increased its crystal structure changes and it is converted first into coesite and finally into stishovite. Coesite is found as minute inclusions in diamond. Recently Joseph R. Smyth of the Los Alamos Scien-

tific Laboratory and C. J. Hatton of the University of Cape Town have also found relatively large crystals of coesite in an eclogite from a South African kimberlite pipe. Stishovite, on the other hand, has never been observed in kimberlitic material. The stishovite-coesite inversion curve appears to cross the continental geotherm at a pressure of about 100 kilobars, corresponding to a depth of about 300 kilometers. Thus the presence of diamond and the absence of stishovite suggest that kimberlites originate at a depth of between 100 and 300 kilometers.

A further check on these findings can be made by comparing the melting point of peridotitic rocks at various pressures and temperatures with the continental geotherm. Since the ultramafic nodules were brought to the surface by volcanic activity, they must clearly have come from a region subject to episodes of melting. The comparison suggests that the mantle is most likely to melt at a depth of very roughly 100 to 200 kilometers, where the melting curve and the geotherm approach each other most closely. A rather modest increase in temperature at this depth would melt dry peridotite. Alternatively the addition of a comparatively small amount of water would depress the peridotite melting point enough to induce at least partial melting.

The peridotite melting curve and the

presence of diamond and coesite in kimberlite suggest a broad range of depths from which all ultramafic nodules might be derived. It would also be useful to estimate the depth of origin for individual nodules. Such estimates can be made by determining the state of mineral systems in the nodule and again calibrating the measurements with laboratory observations.

One useful indicator is the extent to which orthopyroxenes and clinopyroxenes are dissolved in each other. In 1966 B. T. C. Davis and Francis R. Boyd of the Carnegie Institution of Washington's Department of Terrestrial Magnetism showed that the mutual solubility of these two phases increases with temperature but is almost independent of pressure. The solubility is thus a potentially useful geothermometer. Pressure can be determined in a similar way by measuring the extent of reaction between coexisting orthopyroxenes and garnets. In this case the degree of reaction is dependent on both pressure and temperature, but the pressure can be deduced if the temperature is estimated first from the orthopyroxene-clinopyroxene solubility. It is important to note that the temperatures and pressures calculated in this way do not necessarily correspond to the depth where the nodules were originally formed. They give only the depth where the mineral systems were last in equilibrium. Prolonged residence at a level above or below the site of formation could "reset" both the thermometer and the pressure gauge.

The first pressure and temperature data for kimberlite nodules were calculated in 1973 by Boyd and P. H. Nixon. They found that nodules from a single pipe often exhibit a wide range of equilibrium pressures and temperatures. The data are not scattered randomly, however; pressure and temperature are correlated in such a way that the values for most nodules fall in a narrow band. The position of the band was found to vary somewhat from pipe to pipe, but on the whole it coincides with the continental geotherm predicted by heat-flow measurements. What this pattern implies is that the ascending kimberlite incorporates an essentially random sampling of the mantle material through which it passes.

One surprising discovery to emerge from Boyd and Nixon's calculations was that the texture of the nodules varies with their apparent depth of origin. In mineralogy texture refers to the spatial relations of mineral grains. In the most common peridotitic nodules the grains are about half a centimeter across; they show no apparent deformation of the crystalline structure and they give little evidence of having a preferred orientation. In other nodules, however, the structure of some crystals is altered and other crystals are ground up, changes that can be interpreted as the effects of deformation by shearing. The onset of deformation is indicated by olivine crystals, which develop striations, called kink bands, in their crystal lattice and which also become granulated at their margins. With further deformation the rock breaks up into a matrix of finely milled olivine with fragments of orthopyroxene and garnet as insets. Extremes of deformation lead to finely banded rocks in which all the minerals, with the frequent exception of garnet, are much reduced in grain size. In some specimens there is evidence of recrystallization leading to the formation of coarsegrained rocks that differ from ordinary granular peridotites in that the grains have a preferred orientation. The deformed specimens are collectively called sheared nodules.

Several pressure and temperature curves calculated by Boyd and Nixon in 1973 indicated that sheared nodules almost invariably come from a greater depth than granular nodules. Moreover, there appeared to be a sudden upward inflection of the geotherm at the depth of the sheared nodules, that is, the temperature at that depth seemed to be higher than would be expected from heat-flow measurements at the surface. There was widespread speculation that the sheared nodules might represent samples of material from the asthenosphere. The sliding of lithospheric plates over the mantle in this region could certainly lead to extreme deformation of rocks, and the resulting frictional heating might be responsible for the higher temperatures.

This intriguing proposal has since been challenged. The calculation of equilibrium pressures and temperatures has been refined, notably by B. J. Wood and S. Banno of the University of Manchester. Recalculated pressure and temperature curves for kimberlite nodules, such as those made by J.-C. Mercier and Neville L. Carter of the State University of New York at Stony Brook, show no inflection at the depth of the sheared nodules. On the other hand, the observation that sheared nodules generally come from greater depths than granular nodules has been confirmed. The depth estimates are also in general agreement with those made by other techniques. Most equilibrium pressures are in the range from 30 to 80 kilobars, corresponding to depths of roughly 100 to 250 kilometers.

Several workers have proposed alternative explanations for the origin of the sheared nodules, in which the movement of the kimberlite itself is responsible for the deformation. For example, Harry W. Green of the University of California at Davis and Y. Gueguen of

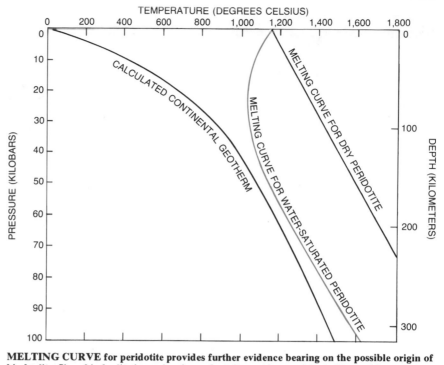

MELTING CURVE for peridotite provides further evidence bearing on the possible origin of kimberlite. Since kimberlite is a volcanic product, it must have originated in a region where at least some mantle materials are molten. The melting curve for dry peridotite (*black*) approaches the continental geotherm most closely in the expected depth range of 100 to 300 kilometers. Melting could result from episodic heating of the mantle at that depth, which would produce an upward inflection in the geotherm. Alternatively the addition of water to the mantle rocks (*color*) could depress the melting curve for peridotite enough to cause at least partial melting.

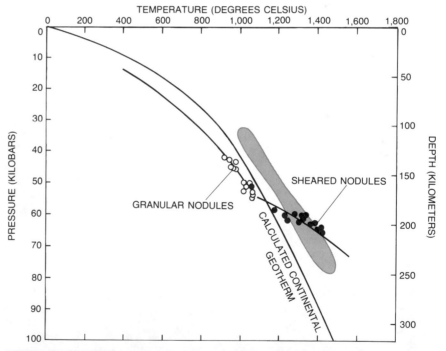

DEPTH OF ORIGIN for ultramafic nodules was estimated by calculating the temperature and pressure at which certain mineral systems in the nodules were last in equilibrium. A series of calculations made in 1973 by Francis R. Boyd and P. H. Nixon of the Carnegie Institution of Washington's Department of Terrestrial Magnetism suggested that the nodules could be divided into two series based on their texture. Granular nodules, the commoner type (*open dots*), fall along a curve near the predicted continental geotherm. Highly deformed specimens called sheared nodules (*solid dots*), however, seem to be derived from deeper levels and also suggest that temperatures there are higher than those calculated from heat-flow measurements. One explanation for this pattern is that the sheared nodules came from the asthenosphere, where the sliding of lithospheric plates could lead to both deformation and heating. Pressures and temperatures for the nodules have been recalculated by Neville L. Carter and J.-C. Mercier of the State University of New York at Stony Brook. Their results (*colored band*) confirm that the sheared nodules come from deeper strata but show no inflection in the geotherm.

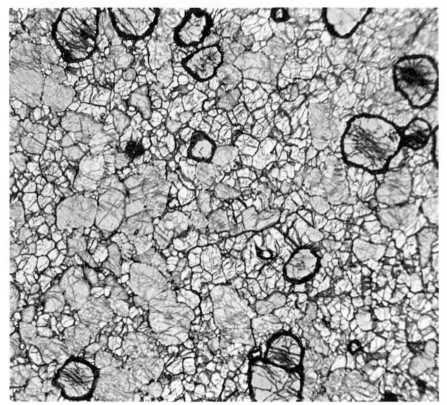

GRANULAR NODULE from the Matsoku pipe in Lesotho reveals an undeformed structure in a micrograph of a polished section. The four major constituents of peridotite are present. The pink (or brown) crystals with dark margins are garnet, green crystals are chrome diopside, the buff-colored crystals are orthopyroxene and the small, colorless crystals are olivine. The micrograph, which was made in natural light, enlarges the specimen to twice its natural size.

SHEARED NODULE is seen in a photomicrograph made between crossed polarizing filters in order to enhance contrast between crystals. The large crystals are orthopyroxene. They are set in a matrix of olivine that has been reduced to a finely milled powder by extreme deformation. The specimen is from the Bultfontein pipe at Kimberley. It is magnified some 20 times.

the Laboratory of Structural Geology in Nantes have proposed a model in which an ascending mass of solid peridotite develops a skin of highly deformed rocks. These rocks are then incorporated as sheared nodules into magma formed by melting near the top of the structure. It must be emphasized, however, that the origin and significance of the sheared nodules remains a subject of research and debate. A consensus has not yet emerged.

Although ultramafic nodules are derived from the mantle, it is by no means certain that they are typical of native mantle material. The primitive mantle, formed in the early stages of the earth's history, was presumably much like the material observed in chondritic meteorites, modified only by a melting episode in which the iron-nickel phase was removed to form the core. It seems unlikely that mantle material of this kind might be identified in peridotite nodules found at the surface today. Indeed, evidence from the ratios of isotopes in various igneous rocks suggests that there have been several major episodes of melting in the upper mantle.

One way in which the upper mantle has been altered is by the continuous removal of the more volatile components through volcanic activity. The material represented by the peridotitic nodules may therefore be a refractory residue. When peridotite is heated, one of the first fractions to melt has a composition like that of basalt, and basalt has erupted voluminously in virtually all geological periods. Basalts have more calcium and aluminum than peridotite has, and so the residual mantle should be depleted of these elements; on the other hand, it should be comparatively rich in magnesium. These alterations would be expressed mineralogically by the rareness or absence of garnet and clinopyroxene, as in the commonest kimberlite nodules. Significantly, these nodules appear to have a comparatively shallow source, and the sheared nodules from deeper layers are usually less depleted in basaltic components.

Some kimberlite nodules might also be derived from basaltic liquids that had been trapped in the mantle and crystallized there as cumulates, which are collections of crystalline phases deposited by such liquids as they migrate to the surface. Several nodules with abundant clinopyroxene and garnet have been tentatively identified as cumulates.

The interpretation of kimberlites is complicated by the eventful history of the upper mantle. Even several hundred kilometers under the surface the composition and crystal structure of rocks are altered repeatedly by a variety of chemical and physical processes. For example, fluids containing dissolved salts can penetrate the grain boundaries and microfractures of solid rock. Chemical re-actions with the dissolved ions can com-pletely change the character of the host rock. Phlogopite, the mica found in pe-ridotitic nodules, is frequently formed in this way.

Melting followed by slow cooling and recrystalization has also probably al-tered the structure of many rocks incor-porated in kimberlite nodules. Much of the evidence required for recognition of their source is thereby destroyed. Nev-ertheless, some events in the history of the nodules can be reconstructed. For example, in many nodules garnet and clinopyroxene tend to be closely associ-ated. In an earlier, high-temperature stage they may have been dissolved in each other or they may both have been dissolved in an orthopyroxene. On cool-ing, the two components were precipi-tated, leaving only the spatial associa-tion as a clue to the earlier structure.

The origin of eclogite nodules has proved particularly difficult to compre-hend. The eclogites are similar to basalt in composition, and they have been widely interpreted as cumulates precipi-tated from basaltic liquids. Their band-ed texture is not always consistent with this hypothesis, however. An earlier al-ternative explanation proposed by Al-fred E. Ringwood of the Australian Na-tional University suggests that they de-rived from sections of oceanic crust drawn into the mantle at subduction zones, where converging continents and seafloor meet. If eclogite nodules do originate in this way the presence of dia-monds in them is particularly intriguing.

The origin of the kimberlite matrix is perhaps even more obscure than that of the ultramafic nodules. Kimber-lites have the high magnesium content typical of mantle liquids, but they are also extremely rich in volatile elements and have higher concentrations of po-tassium than is normal for such magne-sium-rich rocks. Liquids similar to kim-berlite might be produced by the frac-tional crystallization of basaltic mag-mas, and it may not be a coincidence that both southern Africa and Siberia were sites of widespread basaltic vol-canism. Trapped pockets of molten ba-salt could conceivably have evolved toward a kimberlitic constitution, al-though in some cases the fluid must have remained trapped for as long as 100 mil-lion years. Alternatively fluid kimberlite might be the product of very slight par-tial melting of the mantle material, per-haps enriched in certain elements by the migration of fluids through it.

The interpretation of kimberlite and the nodules it contains would surely be more secure if their history were less complicated. Even if the story they tell is for now a confusing one, however, they remain among the best available sources of information about the mate-rial of the upper mantle.

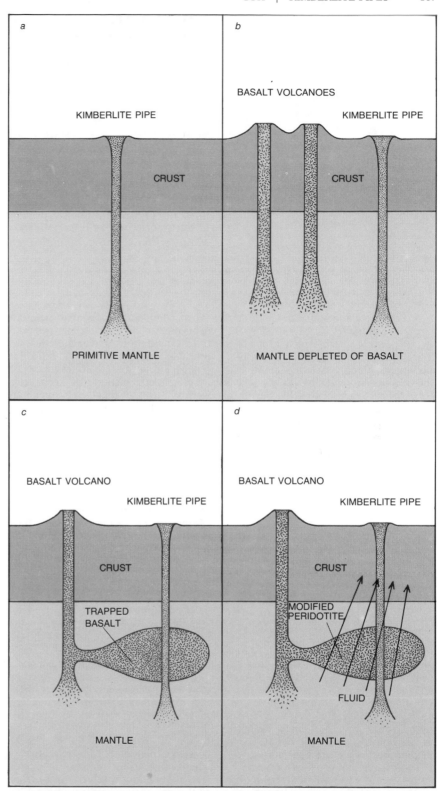

SOURCE MATERIAL of ultramafic nodules is most likely to be some constitutent of the up-per mantle, but its exact nature remains in doubt. One possibility (*a*) is that the nodules are specimens of primitive mantle material, essentially unaltered since the early stages of the earth's history. This model is not considered plausible because many processes that could alter the composition of the mantle have long been operating; for example, the extensive eruption of ba-salt lavas could have left the mantle depleted of certain elements and enriched in others. Anoth-er hypothesis (*b*) is that the nodules represent just such a residue of mantle material depleted of basalt. The expected composition of such material is similar to that observed in the commonest peridotite nodules. Other kinds of nodules may be derived from basalt trapped in the mantle or from cumulates precipitated there by basaltic magma migrating to the surface (*c*). Finally, composition and mineral structure of such trapped materials (and indeed of almost all mantle rocks) could be locally modified by the passage of mineral-bearing fluids through them (*d*).

9

The Chemical Evolution of the Earth's Mantle

by R. K. O'Nions, P. J. Hamilton,
and Norman M. Evensen
May 1980

*Clues to the nature and timing of the mantle's
differentiation are gained from precise measurements
of the isotopic ratios of certain trace elements in rock
samples from continental and oceanic crust*

According to the prevailing plate-tectonic theory of geophysics, the oceanic crust, which represents more than two-thirds of the earth's solid surface, is being constantly created at the mid-ocean ridges by flows of lava from the earth's interior. This veneer of basaltic rock, averaging less than a thousandth of the earth's radius in thickness, rides along on top of the considerably denser mantle for as much as thousands of kilometers to a subduction zone, where it plunges downward, returning to the deep recesses of the mantle. The thicker, lighter continents, which account for the remaining third of the globe's surface area, are literally islands of stability amid this shifting scene. Whereas the oldest-known parts of the oceanic crust are about 200 million years old, an area of the continental crust has been found to be close to 3.8 billion years old: almost 20 times older. Yet old as the continents are, they are apparently not primordial features of the earth but rather secondary features that have formed and evolved during the earth's lifetime. Indeed, they appear to be accreting even today. Recent advances in geochemistry offer a clearer view of how the continents have arisen out of material of the mantle and, by inference, how and when the earth as a whole became differentiated into its present multilayered structure.

The continental crust has a chemical composition very different from that of the rest of the earth. Although the continents make up only .4 percent of the earth's mass, they have concentrated within them disproportionately large shares of the earth's inventory of certain elements. Among the most notable of these elements are the heat-producing radioactive isotopes of potassium, thorium and uranium; a significant fraction of each of these elements now resides in the continents. Within the past two decades it has been demonstrated on the basis of the systematic radiometric dating of continental rocks that the continents themselves must have grown more or less continuously over the course of geologic time. In the early 1960's Patrick M. Hurley and his colleagues at the Massachusetts Institute of Technology were able to show that the continents are the product of many separate episodes of chemical differentiation, and a great deal of evidence has since been adduced in support of this conclusion [see "The Oldest Rocks and the Growth of Continents," by Stephen Moorbath; SCIENTIFIC AMERICAN Offprint #357].

The outstanding questions concerning the role of the earth's mantle in producing the continents include the following: How much of the mantle is involved in the formation of the continents? Do the mantle-derived lavas that have erupted at the surface come from previously differentiated mantle or from undifferentiated mantle? Over what part of the earth's history have the continents developed and at what rate? How much early continental material has been recycled back into the mantle?

The search for satisfactory answers to these questions involves the difficult task of deciphering from a very incomplete geologic record the comparative distributions of various elements between the mantle and the continental crust as a function of time. In a situation where much of the material of interest is securely out of reach one must resort to the use of geochemical "tracers" to reconstruct some record of the chemical fractionation and differentiation that has taken place. The most recent and

successful work of this kind has exploited the nonradioactive decay products of radioactive trace elements with half-lives comparable in duration to the age of the earth. These naturally occurring "parent-daughter" associations include the parent isotopes potassium 40, rubidium 87, samarium 147, thorium 232, uranium 235 and uranium 238, together with their respective daughter isotopes calcium 40 (or argon 40 in the case of electron capture by potassium 40), strontium 87, neodymium 143, lead 208, lead 207 and lead 206 [see illustration on page 110].

One feature common to all these elements is that their ionic radii are much larger than those of the most abundant constituents of the mantle, such as magnesium, aluminum, silicon and iron. The large ionic radii of the trace elements, together with their propensity to substitute for other ions in the comparatively open silicate structures of the rocks of the earth's crust, has led geochemists to refer to them as large-ion lithophile elements. In spite of the comparatively low abundance of the large-ion lithophile elements in the earth, three of them—potassium, thorium and uranium—are responsible for most of the earth's internal heat production. The degree to which these trace elements are concentrated selectively in the continental crust can be appreciated by comparing their abundances in the continental crust with their abundances in the earth as a whole [see top illustration on page 111]. Because the trace elements have larger ionic radii than magnesium, silicon or iron, which make up most of the dense rocks

MATERIAL FROM THE MANTLE is exposed in the form of newly created oceanic crust in this Landsat image of southern Iceland. The entire island is part of the Mid-Atlantic Ridge system. Here the crest of the ridge, an intensely active volcanic zone, runs southwest-northeast. In this false-color image, made in the summertime, the dark gray areas correspond to the most recent lava flows and the red areas are somewhat older lava beds covered with vegetation. Large white shapes are glaciers. Circular white spots are snow-capped volcanoes. The small islands at the bottom include Heimaey and Surtsey, both sites of volcanic eruptions in recent years.

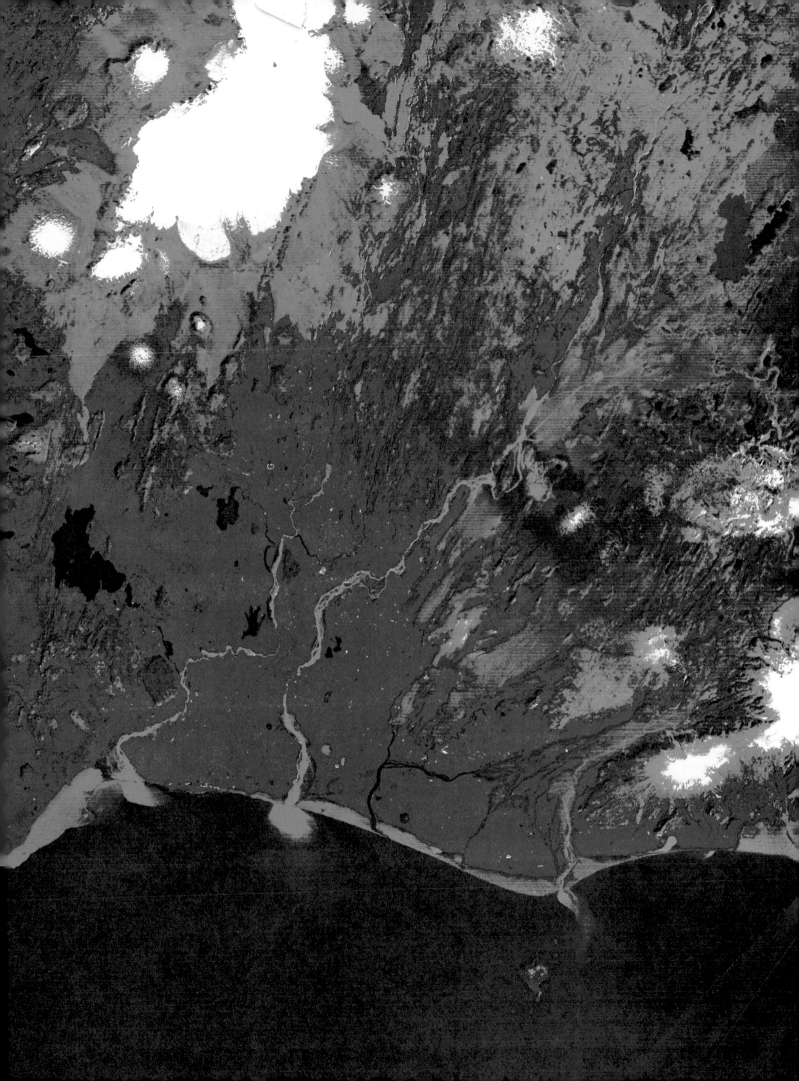

of the mantle, they do not substitute readily in these rocks.

How does the abundance of a particular daughter isotope reflect the fractionation history of a part of the earth? Consider the abundance of the daughter isotope lead 206 at a given time and place within the earth. Some of the lead 206 will have existed from the time of the earth's formation, but it will have been augmented by additions from the decay of the parent isotope uranium 238. The exact amount of lead 206 present will reflect changes in the ratio of uranium to lead as a function of time. If, for example, the fractionation history of uranium and lead is such that at some time in the past one part of the earth had a high uranium/lead ratio and another part had a low uranium/lead ratio, then after a certain elapsed time the relative abundance of lead 206 in the first part would be expected to be greater than that in the second. The difference is conveniently illustrated by expressing the abundance of lead 206 with respect to the abundance of an isotope of lead that has not received radiogenic additions, namely lead 204. Thus the fractionation history of uranium and lead in the earth is recorded in the variability of the lead 206/lead 204 ratio from time to time and place to place.

In a similar way the ratios of lead 208 to lead 204, of lead 207 to lead 204, of neodymium 143 to neodymium 144, of strontium 87 to strontium 86 and of argon 40 to argon 36 will respectively reflect the fractionation histories of the thorium-lead, uranium-lead, samarium-neodymium, rubidium-strontium and potassium-argon parent-daughter associations. Because the half-lives of the parent isotopes are all comparable to the age of the earth the present isotopic ratios of argon, strontium, neodymium and lead can be taken as a reflection of parent-daughter fractionations that have taken place over the entire history of the earth. The daughter of a parent isotope with a much shorter half-life, such as xenon 129, the decay product of iodine 129, which has a half-life of 16 million years, can reflect only the fractionation of iodine and xenon that occurred in the first couple of hundred million years after the formation of the earth. Parent-daughter associations with such short half-lives are of limited value for studying processes that have occupied a large fraction of the earth's history, such as the growth of continents.

The selective removal of large-ion lithophile elements from the mantle and their concentration in the developing continental crust are processes that involve relative fractionations in the abundances of these elements. For example, the continental crust develops with a higher ratio of rubidium to strontium and a lower ratio of samarium to neodymium than exist in the mantle;

in both cases the continents selectively concentrate the element with the larger ionic radius of the pair. As a result the continents evolve with a greater relative abundance of strontium 87 and a lesser relative abundance of neodymium 143 than the mantle, a phenomenon that is recorded in a higher strontium 87/strontium 86 ratio and a lower neodymium 143/neodymium 144 ratio in the continents compared with the residual mantle. In principle, therefore, isotopic studies of new additions of material to the continents at various times can provide information about the previous differentiation history of the source from which the sample was extracted. Similarly, the present isotopic composition of mantle-derived samples will in principle yield information about the mantle's prehistory.

This article is primarily concerned with recent developments in geochemistry that furnish insights into the nature and timing of the processes that have produced the earth's present chemically differentiated character. Before

PARENT ISOTOPE	DAUGHTER ISOTOPE	OTHER DECAY PRODUCTS	HALF-LIFE (BILLIONS OF YEARS)
RUBIDIUM 87	STRONTIUM 87	+ 1 ELECTRON	48.8
POTASSIUM 40	CALCIUM 40	+ 1 ELECTRON	1.47
POTASSIUM 40 + 1 ELECTRON	ARGON 40		11.8
URANIUM 238	LEAD 206	+ 8 ALPHA PARTICLES, 6 ELECTRONS	4.468
URANIUM 235	LEAD 207	+ 7 ALPHA PARTICLES, 4 ELECTRONS	.7038
THORIUM 232	LEAD 208	+ 6 ALPHA PARTICLES, 4 ELECTRONS	14.008
SAMARIUM 147	NEODYMIUM 143	+ 1 ALPHA PARTICLE	106

LARGE IONIC RADII of an assortment of radioactive trace elements and their respective major decay products are crucial to the effectiveness of these isotopes as geochemical probes. The ions constituting each "parent-daughter" association are drawn to the same scale at the left. (The smallest of these isotopes, samarium 147, has an ionic radius of .964 angstrom unit.) As the column at the right shows, each of the parent isotopes has a half-life on the order of (or greater than) the estimated age of the earth (4.55 billion years). Potassium 40 can decay by either capturing or emitting an electron, leading to different daughter isotopes. With the exception of calcium, all the parent and daughter isotopes shown are present in trace amounts in the earth (one part per million or less). Each of the trace elements has an ionic radius larger than that of the much commoner elements silicon, aluminum, magnesium, calcium and iron, which in the form of oxides make up most of the earth's mantle. The large ions do not fit into the dense crystal structures of the mantle minerals, which generally accommodate only the more abundant smaller ions. As a result the larger ions migrate to the crust, where they reside in less dense crystal structures. The largest ion, that of argon 40, can escape into the atmosphere, which is about 1 percent argon. This argon is almost entirely the product of the decay of potassium 40.

getting into the details it will be helpful to compare the abundances of some of the elements in the earth that are relevant to our story with their estimated abundances in the primordial solar nebula and also in the moon, the only other sizable object in the solar system for which appropriate data are currently available.

The inner, terrestrial planets, such as the earth and Mars, have in general retained a much smaller complement of the most volatile elements (hydrogen, for example) from the solar nebula than the outer, giant planets, such as Jupiter and Saturn. Furthermore, recent planetary explorations have uncovered significant differences in composition among the earth, Mars and the moon. Much of the ensuing discussion will be concerned with the abundances and isotopic ratios of potassium, argon, rubidium, strontium, neodymium, samarium, lead, thorium and uranium, all of which are trace constituents not only of the earth but also of the other terrestrial planets. For comparative purposes magnesium, aluminum and silicon, which are major constituents of both the earth and the other terrestrial planets, will also enter the discussion. As far as these particular elements are concerned the meteorites known as carbonaceous chondrites are considered to represent the best approximation of their abundances in the primordial solar nebula.

The first point to be noted from a comparison of this type is that the abundances of most of the elements named are roughly the same in the earth, the moon and the carbonaceous chondrites. In contrast, potassium, rubidium and lead are apparently less abundant in the earth and the moon than in the chondrites, the extent of the depletion being greater in the case of the moon. The relative depletion of potassium can be demonstrated on the basis of the measured potassium/uranium ratios of terrestrial, lunar and chondritic samples. The relative depletion of rubidium and lead can be demonstrated on the basis of the known isotopic compositions of the strontium and lead in these three materials.

Lawrence Grossman of the University of Chicago has calculated that in a gas of the same composition as that of the sun calcium, aluminum, strontium, neodymium, samarium, uranium and thorium would all be quite refractory, condensing at temperatures higher than 1,350 degrees Kelvin, whereas potassium, rubidium and lead would all condense at lower temperatures, the least refractory of these being lead, which would condense at about 520 degrees K. It is beyond the scope of this article to review the possible mechanisms by which this fractionation of elements of differing volatility could have occurred in the earliest stages of the solar system; it is sufficient here merely to note that

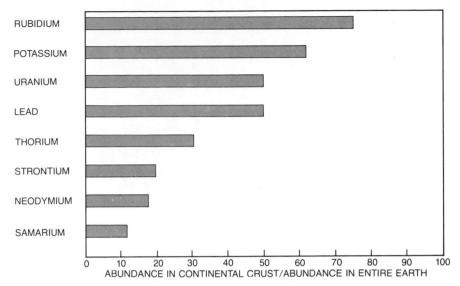

ABUNDANCES OF SOME LARGE-ION TRACE ELEMENTS in the continental crust, estimated by S. R. Taylor of the Australian National University, are given as a function of their estimated abundances in the earth as a whole. The elements with the largest ionic radius tend to exhibit the greatest degree of enrichment in the crust. In spite of the insignificant mass of the crust compared with that of the mantle, a significant fraction of the earth's total inventory of these large-ionic-radius elements resides in the crust. Because of the propensity of such elements to substitute for those with smaller ionic radii in the comparatively open silicate structures of the crustal rocks, they are referred to by geochemists as large-ion lithophile elements.

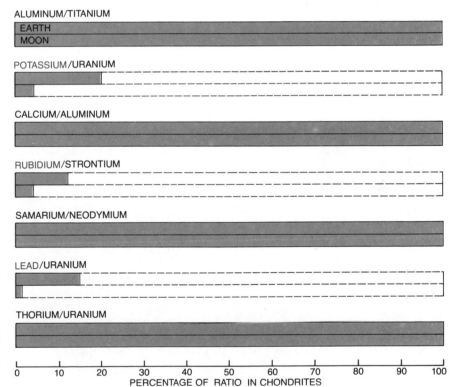

THE EARTH AND THE MOON APPEAR TO BE DEPLETED in certain volatile trace elements (*colored type*) when the ratios of pairs of elements are compared with corresponding ratios in carbonaceous chondrites, a class of meteorites that are considered to be representative of the average composition of the solar system. (Ratios are used for this purpose rather than average abundances, because they can be determined with greater precision.) When both elements are refractory, and hence would presumably have condensed together at high temperatures out of the primordial solar nebula, their ratio is identical in the earth, the moon and the chondrites. The ratio of volatile elements to refractory ones is found to be lower in the earth than in the chondrites, however, and it is lower still in the moon. The volatile elements would have condensed out of the cooling solar nebula later than the refractory ones; the earth and the moon appear to be depleted in this volatile fraction. (Here lead refers to primordial lead and does not include lead produced since the earth's formation by decay of uranium and thorium.)

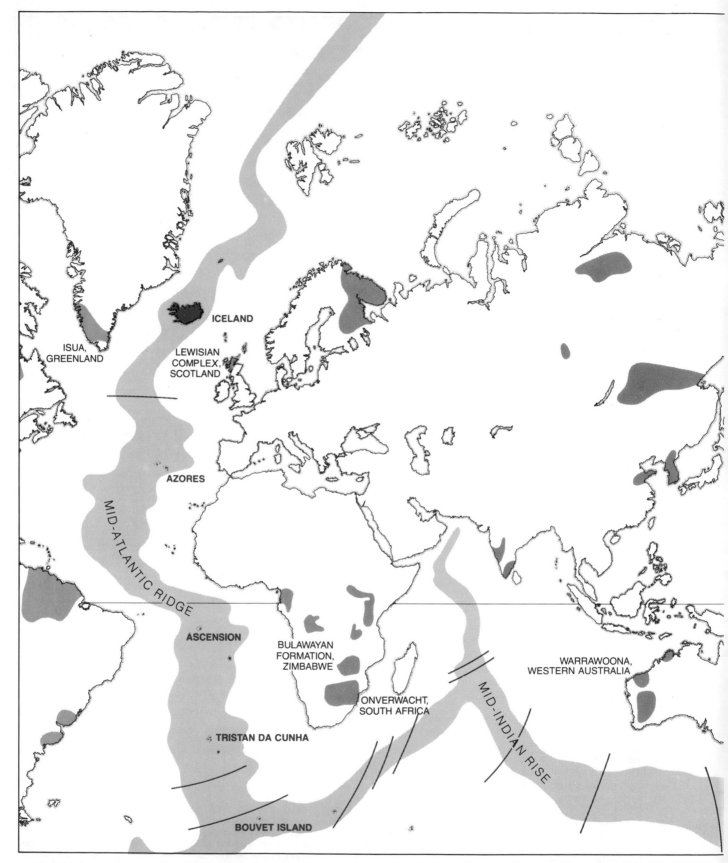

ROCK SAMPLES analyzed in the investigation of the isotopic ratios of large-ion lithophile elements in the continental crust and the oceanic crust are located. Parts of the continental crust more than 2.5 billion years old are in gray; in general they are found in the most ancient, tectonically stable "shield" regions of the continents. Rocks from the named regions have figured specifically in the recent determination of the early isotopic history of the earth. The colored areas include the youngest parts of the earth's crust: the mid-ocean ridges, where new crust is created by the flow of lava from the mantle. Isotopic analyses of these lavas indicate they are derived from mantle

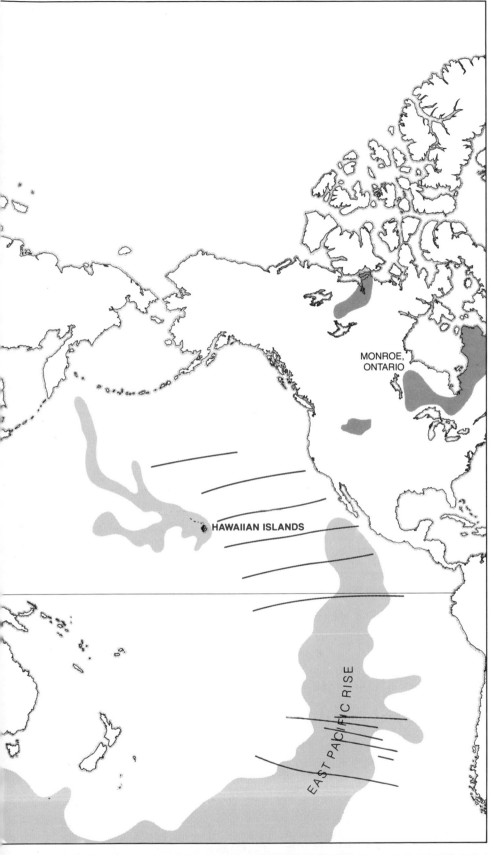

MONROE,
ONTARIO

HAWAIIAN ISLANDS

EAST PACIFIC RISE

material that is depleted in the very elements (particularly in the heat-producing radioactive trace elements potassium, thorium and uranium) selectively concentrated in the continental crust. The oceanic islands, marked in solid color, have emerged as a result of particularly voluminous flows of lava. Analyses of rocks from these islands suggest that the rocks are derived from mantle material less depleted than that responsible for the rest of the mid-ocean ridges.

the earth and the moon appear to have formed with an assortment of refractory elements present in roughly chondritic proportions, but with a marked (and in the case of the moon an extreme) depletion in the more volatile elements. (The notion that the earth has a chondritic composition was popular for many years among geophysicists investigating terrestrial heat flow because of the coincidence of the conductive heat loss from the earth and the average rate of heat production in the chondritic meteorites. It now appears, however, that the ratios of the refractory elements thorium and uranium, two of the main heat-producing elements in the earth, to the more volatile potassium, the third important heat-producing element, in the earth differ greatly from those in the chondrites.)

Having considered briefly the abundances of some elements in the entire earth in the context of the composition of the primordial solar nebula, let us now turn our attention specifically to the differentiation of the earth itself. We shall first consider the progress that has been made in extracting isotopic data from some of the older parts of the continents. The prime objective of such investigations is to derive the values of the strontium 87/strontium 86, neodymium 143/neodymium 144, lead 208/lead 204, lead 207/lead 204 and lead 206/lead 204 ratios for the source region of a given segment of continental crust and to compare these values with the predicted isotopic ratios for undifferentiated mantle at that time.

Since 1975 considerable progress has been made in exploiting the samarium-neodymium system for the solution of problems in geochronology and in utilizing neodymium 143 as a natural tracer of geological processes. The rubidium-strontium and uranium-lead systems have been exploited successfully for two decades or more, but the application of the samarium-neodymium system was impeded by technical difficulties involved in the extraction of samarium and neodymium from rock samples and in the isotopic analysis of these elements to the required degree of precision. Because samarium 147 decays very slowly, extremely small differences in the abundance of neodymium 143 must be determined. Precise mass-spectrometric techniques now exist for such measurements, largely because of the pioneering efforts of Gerald J. Wasserburg and his colleagues at the California Institute of Technology. In 1975 Guenter W. Lugmair of the University of California at San Diego published the first precise isotopic analyses of the samarium-neodymium system in an achondritic meteorite and a lunar sample. Subsequently C. J. Allègre and his colleagues at the Univer-

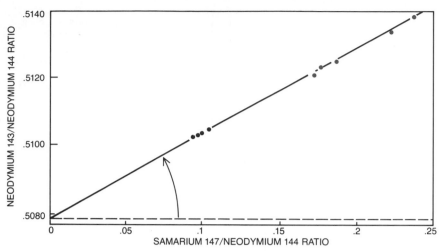

ROCKS FROM ISUA, an ancient metamorphosed volcanic deposit in western Greenland, are dated by means of measurements of their samarium/neodymium and neodymium-isotope ratios. Assuming that the rocks formed originally from a part of the mantle with a uniform ratio of neodymium 143, the daughter isotope of samarium 147, to neodymium 144, an isotope that is neither generated nor consumed by radioactive decay, then the neodymium 143/neodymium 144 ratios of the rocks will increase with time, at rates that are proportional to their samarium 147/neodymium 144 ratios. Basic igneous rocks (*colored dots*) form with higher samarium 147/neodymium 144 ratios, closer to the original mantle value, than the more acid, silica-enriched rocks (*black dots*). In a plot of neodymium 143/neodymium 144 against samarium 147/neodymium 144, such as this one, all the dots would originally lie on the broken horizontal line. As the rocks evolve and develop higher neodymium 143/neodymium 144 ratios with time, the faster growth of the neodymium 143/neodymium 144 ratio in rocks with higher samarium 147/neodymium 144 ratios will cause the line to pivot upward at the right. The slope of the resulting line, called an isochron, indicates how long the rocks have followed their separate evolutionary paths since forming from a homogeneous reservoir. The age of the Isua samples determined by this method is 3.77 billion years, more than four-fifths of the age of the earth. Any disturbance that added or removed samarium or neodymium during that time would tend to destroy the linearity of the isochron. Only a hypothetical rock with a samarium 147/neodymium 144 ratio of zero would display no change in its neodymium 143/neodymium 144 ratio and so would preserve the original value of this ratio at the time the rocks formed. Since rocks with samarium 147/neodymium 144 ratios much lower than those shown here are not likely to form, the initial neodymium 143/neodymium 144 ratio of the Isua rocks is obtained by extrapolating the isochron to zero on the samarium 147/neodymium 144 axis. The initial ratio found in this way is an indicator of the isotopic composition of the mantle at that time (3.77 billion years ago). Similar measurements can be made with other parent-daughter associations, but the proportions of most such elements change in the course of metamorphism and therefore do not yield as accurate dates or initial isotopic ratios by the isochron technique.

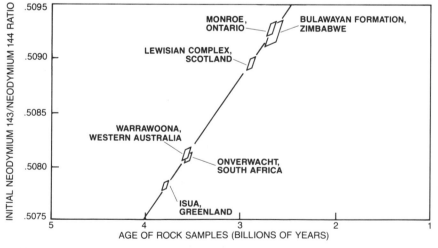

EVOLUTION OF THE RATIO of neodymium 143 to neodymium 144 in the mantle can be followed by plotting the ages of rock samples from six ancient regions of the continental crust against their initial neodymium 143/neodymium 144 ratios, derived by the isochron technique. The symbols in this case are parallelograms within which there is a 95 percent probability that the true value lies. All the age determinations summarized here were obtained by the authors, except for the dating of the rocks from Monroe Township in Ontario, which was carried out by Alan Zindler and Stanley R. Hart of the Massachusetts Institute of Technology. The fact that the points lie close to a line representing the growth of the neodymium 143/neodymium 144 ratio in meteorites with a characteristically chondritic samarium 147/neodymium 144 ratio is evidence that the mantle formed with a chondritic samarium/neodymium ratio.

sity of Paris, Donald J. DePaolo and Wasserburg at Cal Tech, and our group, which was then at the Lamont-Doherty Geological Observatory of Columbia University, published the results of neodymium-isotope studies for a range of terrestrial samples and demonstrated the utility of the neodymium-isotope approach for investigating a variety of geological problems.

The samarium-neodymium system has been extremely valuable in studying the oldest components of the continental crust because of its survival through various alteration processes. The ratio of rubidium to strontium is often disturbed by weathering and other processes that affect volcanic rocks after their eruption, making the observed strontium 87/strontium 86 ratios difficult to interpret. In contrast, the work done so far has shown that the samarium/neodymium ratios are much less likely to be disturbed. This fortunate attribute of samarium and neodymium has enabled us to obtain precise samarium/neodymium ages and initial neodymium 143/neodymium 144 ratios for some old parts of the continents [*see top illustration at left*]. In particular we obtained samarium/neodymium ages for some supracrustal rocks (that is, rocks deposited on preexisting crust) discovered at Isua in Greenland that are among the oldest-known rocks on the earth. Although the samples we studied were altered considerably since their original extrusion as lava, they have yielded a precise age of 3.77 billion years, which is in excellent agreement with the date obtained by Allègre's group in Paris, based on the analysis of the uranium/lead ratio in samples of zircon (zirconium silicate) from the Isua site. In addition to determining the time the Isua samples formed we also established the neodymium-isotope ratio of their source region with high precision. Precise samarium/neodymium ages and initial neodymium-isotope ratios have also been obtained by the authors from the Onverwacht group in South Africa (3.54 billion years), the Warrawoona group in Western Australia (3.56 billion years), the Lewisian complex in Scotland (2.92 billion years) and the Bulawayan formation in Zimbabwe (2.64 billion years).

The ages and the initial neodymium-isotope ratios of these rocks may be plotted against one another and compared with the evolution of the neodymium-isotope ratio from 4.55 billion years ago in material that has a chondritic samarium/neodymium ratio, as deduced from the analysis of meteorites. One obtains a line corresponding to a samarium/neodymium ratio of .31, which is the current best estimate of the cosmic-abundance ratio. The fact that the data for terrestrial samples more than 2.5 billion years old plot close to this line indicates that the source region of early crust and by inference the earth as a whole had a samarium/neodymium ratio indistinguishable from the cosmic-abundance ratio. The separation of continental crust leaves the residual mantle with a higher ratio of samarium to neodymium than the earth as a whole; hence the average neodymium-isotope ratio of the present mantle cannot lie on the extrapolated portion of the line but must lie above it. Furthermore, the results obtained are consistent with the apparent absence of continental rocks older than the Isua samples. The separation of large amounts of continent before about 3.8 billion years ago, if it had occurred, would be identifiable from the initial neodymium-isotope ratios of basic volcanic rocks erupted in more recent times and preserved in the continents.

Armed with the knowledge that undifferentiated mantle should always have had a ratio of samarium to neodymium equal to the cosmic-abundance ratio, we have been able to make a precise estimate of the present isotopic composition of neodymium in undifferentiated mantle. For the purpose of identifying those parts of the mantle that have donated material to the growth of the continents a survey of the neodymium-isotope composition of the upper mantle would clearly be of considerable value. With the exception of xenolithic (literally "foreign rock") fragments of mantle that are occasionally brought to the surface by kimberlite pipes and some basalt formations, direct sampling of the upper mantle is impossible. As a result it has been necessary to glean information from differentiation products of the mantle, namely the basaltic lavas erupted in the ocean basins and continents. Most recent basalts have been erupted along the mid-ocean ridges, which are the sites of plate generation and sea-floor spreading, but basalts are also erupted in smaller quantities in intraplate locations and along island arcs.

Until the start of the Deep Sea Drilling Project only a comparatively small number of basalt samples had been recovered from the ocean floor, but over the past decade the situation has changed dramatically. The Deep Sea Drilling Project has now entered an international phase of ocean drilling with the participation of France, West Germany, Japan, Britain and the U.S.S.R., and samples have been recovered from a large number of sites in the Atlantic, Indian and Pacific ocean basins. Many additional dredge samples have been obtained from mid-ocean ridges. Probably the most noteworthy of these dredging operations is the systematic and closely spaced dredging along the northern Mid-Atlantic Ridge organized by Jean-Guy Schilling of the University of Rhode Island. In addition to

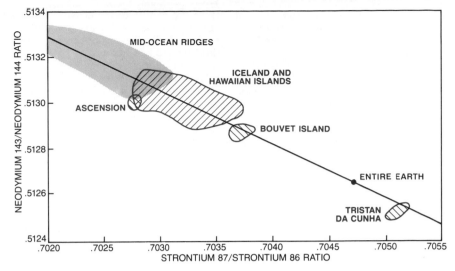

NEGATIVE CORRELATION is observed when measurements of the neodymium-isotope ratio are plotted against measurements of the strontium-isotope ratio for young basaltic rock samples obtained from mid-ocean ridges (*light color*) and oceanic islands (*dark color*). During the formation of the continental crust the mantle was more depleted in rubidium than in strontium and in neodymium than in samarium. Subsequently the neodymium 143/neodymium 144 ratio increased more rapidly and the strontium 87/strontium 86 ratio increased less rapidly in depleted mantle than in undepleted mantle. The mantle underlying the ridges was apparently more depleted than that from which the oceanic islands formed. Assuming that undepleted mantle has a samarium/neodymium ratio characteristic of chondrites, one can calculate the present neodymium 143/neodymium 144 ratio of the entire earth. If undepleted mantle lies on the anticorrelation line, its present strontium 87/strontium 86 ratio and therefore its present rubidium/strontium ratio can also be determined. As depleted mantle evolved chemically away from the point for the entire earth the continents evolved in a complementary manner. The point for typical continental crust would lie off the bottom right-hand corner of the graph.

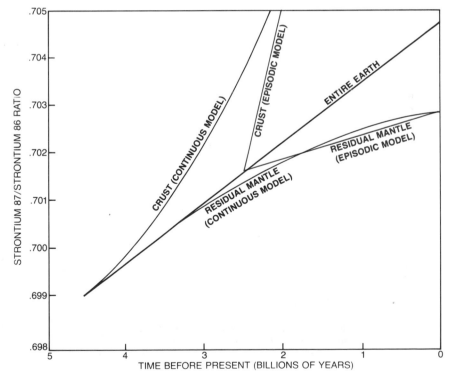

TWO MODELS OF CRUSTAL EVOLUTION are compared in terms of the evolution of their strontium 87/strontium 86 ratios in the crust and the mantle. In the continuous model crust has been produced continuously from the time of the earth's formation, and the crust and the mantle evolve with strontium-isotope ratios that are increasingly different from the ratio of the earth as a whole. In the episodic model it is assumed that the present crust formed 2.5 billion years ago. In actuality older regions of crust are known. Even for this extreme case the evolution of the mantle is quite similar in the two models, since the mantle is a much larger reservoir of strontium than the crust. Curves representing the two models in terms of the evolution of their neodymium-isotope ratios would be even more similar, since samarium and neodymium are less strongly fractionated between the crust and the mantle than rubidium and strontium.

the advances made in ocean-bottom sampling, oceanic islands, particularly Iceland and the Hawaiian Islands, have been further sampled.

The most important advances in understanding the prehistory of the source regions of oceanic basalts have again come from isotopic analyses of strontium and neodymium made over the past few years. The analyses obtained for oceanic basalts exhibit a small range of strontium 87/strontium 86 and neodymium 143/neodymium 144 ratios and are compared with the current best estimates of the same isotopic ratios in the earth as a whole (or in undifferentiated mantle). The ocean-ridge basalts from the Atlantic, Indian and Pacific oceans have isotopic ratios of strontium and neodymium that are respectively lower and higher than the values for the earth as a whole. The lower strontium-isotope ratio requires that the source region of the ocean-ridge basalts have evolved with a lower ratio of rubidium to strontium than the earth as a whole or undifferentiated mantle, whereas the higher neodymium-isotope ratio requires that the samarium/neodymium ratio in the source region be higher than it is in the earth as a whole.

The movement of certain elements from the mantle to the crust has influenced the subsequent isotopic evolution of the mantle. The strontium 87/strontium 86 and neodymium 143/neodymium 144 ratios of recent mid-ocean-ridge basalts seem to be entirely consistent with the hypothesis that components with a higher rubidium/strontium ratio and a lower samarium/neodymium ratio than the earth as a whole have been removed and now reside in the continents. It is not possible to specify unequivocally from these data alone, however, whether the depletion of the mantle has been a continuous phenomenon or an episodic one. Nevertheless, it is clear that the ocean floor created at mid-ocean ridges is derived from a mantle that bears in its isotopic composition the evidence of previous depletion in large-ion lithophile elements.

The comparative uniformity of the isotopic composition of strontium and neodymium in mid-ocean-ridge basalts contrasts with the greater variability of these ratios in oceanic-island basalts. In some instances, such as the volcanic island Tristan da Cunha, the basalts have strontium 87/strontium 86 and neodymium 143/neodymium 144 ratios that are very close to our estimates of the values for the earth as a whole, implying that their source has not evolved in the same way that the sources of the mid-ocean-ridge basalts and may have suffered far less depletion in crustal constituents in the past. It should be emphasized, however, that the volume of volcanic rock erupted on Tristan da Cunha is quite small compared with that erupted at the mid-ocean ridges. Elsewhere in the ocean basins basalts from islands such as Iceland and the Hawaiian Islands have strontium- and neodymium-isotope compositions that are intermediate and overlap those of the mid-ocean ridges.

In short, there is a strong negative correlation between the strontium- and neodymium-isotope compositions of oceanic basalts, indicating a general coherence of parent/daughter ratios during the extraction of the components now residing in the continental crust. Because the samarium/neodymium ratio and therefore the neodymium 143/neodymium 144 ratio of the earth as a whole at present are reliably established the correlation of the strontium- and neodymium-isotope ratios enables us to estimate the strontium 87/strontium 86 ratio in the earth as a whole. Comparison of measured isotopic ratios of neodymium and strontium in volcanic

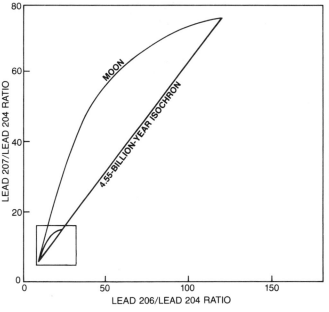

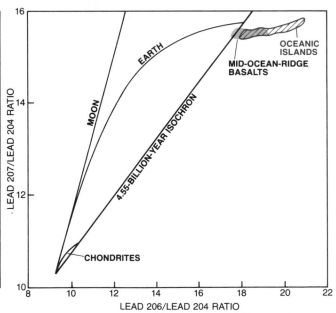

EVOLUTION OF LEAD in the mantle is plotted in the graph at the left and the enlargement at the right in terms of the relative abundances of two isotopes, lead 206 and lead 207, that have been produced at different rates during the earth's history (because of the different proportions and half-lives of the two uranium isotopes that generate them). The two axes of this graph relate the ratios of the two radiogenic lead isotopes to the ratio of lead 204, an isotope whose abundance does not vary with time. On the assumption that a reservoir of homogeneous lead, plotting as a single point on the diagram, was split into several portions with differing ratios of uranium to lead, the lead composition of each portion would evolve with time toward the upper right of the diagram, since both ratios would be increasing. The evolution would be along a growth curve that would depend on the uranium/lead ratio of each portion. At any given time the positions of all the portions along their respective growth curves would lie on a straight line passing through the original lead composition. The slope of the line would be related to the age of the system. In this case the original composition is taken to be that of primordial lead, as measured in iron meteorites with very low uranium/lead ratios. If this ratio was uniform throughout the primordial solar nebula, it would be the point of origin for the growth curves for the earth, the moon and the chondrites, whose uranium/lead ratios differ widely. At present the compositions of these bodies would plot along a line passing through the primordial-lead point with a slope corresponding to the age of the solar system. The compositions of oceanic basalts plot to the right of this line, called the geochron. Samples that appear from an analysis of their strontium- and neodymium-isotope ratios to have come from the most nearly undepleted mantle plot farthest from the geochron in the lead/lead diagram, indicating the evolution of lead in mantle is more complex than that of strontium and neodymium.

rocks with their predicted values for the earth as a whole provides a fine tool for the recognition of depleted-mantle source regions. The variability of isotopic compositions in oceanic basalts bears witness to the fact that the continental crust has not been extracted uniformly, and that the mantle, or at least the part of it under the oceans and efficiently sampled by mid-ocean-ridge volcanism, is depleted in large-ion lithophile elements that now reside in the continental crust.

Evidence for this depletion process was deduced by the late Paul W. Gast in 1968 from considerations of the abundances of trace elements in oceanic basalts. The heterogeneous nature of the upper mantle is also evident from lead-isotope measurements. When these measurements are compared with the predicted average lead-isotope compositions of carbonaceous chondrites, the earth and the moon (assuming that the earth, the moon and meteorites all had the same lead-isotope composition 4.55 billion years ago), one finds that at present the carbonaceous chondrites and other meteorites plot close to a line with a slope corresponding to an age of 4.55 billion years, and one would expect that the bulk earth and moon should also plot close to this line. The relative positions of carbonaceous chondrites, the earth and the moon along this line indicate that the uranium/lead ratio for the moon is greater than the uranium/lead ratio for the earth is greater than the uranium/lead ratio for carbonaceous chondrites, which is consistent with the observation of lower abundances of relatively volatile lead compared with refractory uranium in the earth and the moon.

The lead-isotope compositions of oceanic basalts plot to the right of the 4.55-billion-year line, demonstrating that the suboceanic mantle is inhomogeneous with respect to the isotopic ratios of lead as well as those of strontium and neodymium. Oceanic-island basalts exhibit a greater range of lead-isotope compositions than mid-ocean-ridge basalts, as is also true of their strontium- and neodymium-isotope compositions. At present the existing data for lead, strontium and neodymium isotopes in oceanic basalts are inadequate for a full evaluation of the relations among them. It does seem, however, that a somewhat more complex model must be invoked in order to accommodate the known variations in the lead-isotope ratios. Whereas we have discussed the isotopic data for strontium and neodymium in terms of two reservoirs (the mantle and the continental crust), three reservoirs appear to be required to explain the variations in the lead-isotope ratio.

One of our primary goals in exploiting naturally occurring isotopes as tracers in continental- and mantle-derived

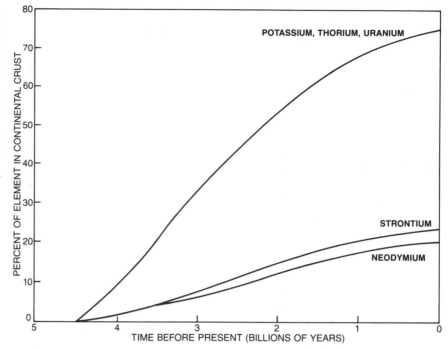

SELECTIVE CONCENTRATION of several large-ion lithophile elements originally present in the mantle but now residing in the continental crust is shown as a function of time. The curves are derived from calculations in which only half of the mantle is assumed to generate continental crust. The vertical axis gives the percent of the original inventory of each element that is present in the crust at any given stage of the earth's history. Heat-producing elements potassium, thorium and uranium have evidently been efficiently extracted from the mantle.

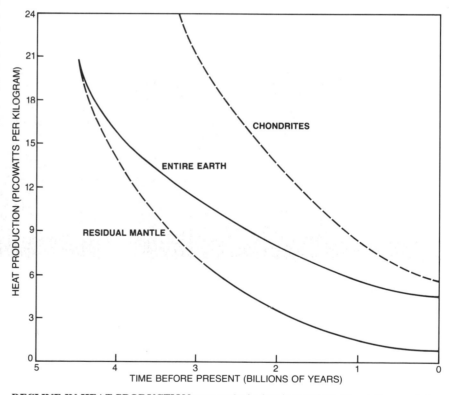

DECLINE IN HEAT PRODUCTION over geologic time is traced for the earth as a whole, the residual mantle (the mantle remaining after the formation of the continents) and the chondrites. All the curves are declining, since the abundances of the heat-producing elements decrease through the same processes of radioactive decay that produce the heat. The heat production in the residual mantle lags increasingly behind that in the earth as a whole as the continents are progressively extracted from the mantle. Heat production in the chondrites is initially high, but it falls off more rapidly than in the earth. This phenomenon is attributable to the fact that chondrites have a higher ratio of volatile potassium to refractory uranium than the earth has; heat from potassium is produced by the radioactive decay of potassium 40, which has a shorter half-life than uranium 238, the isotope that gives rise to most of the heat from uranium.

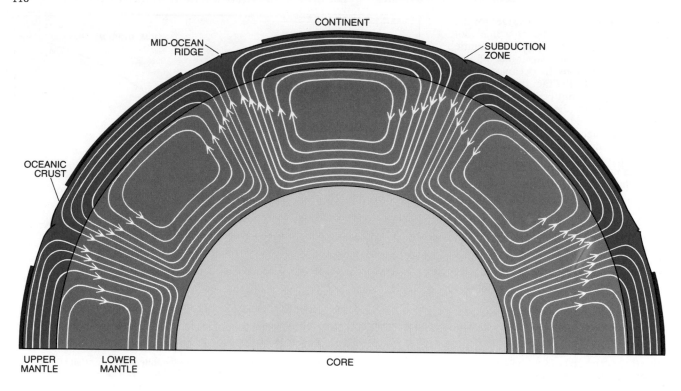

CONTINENT

MID-OCEAN
RIDGE

SUBDUCTION
ZONE

OCEANIC
CRUST

UPPER
MANTLE

LOWER
MANTLE

CORE

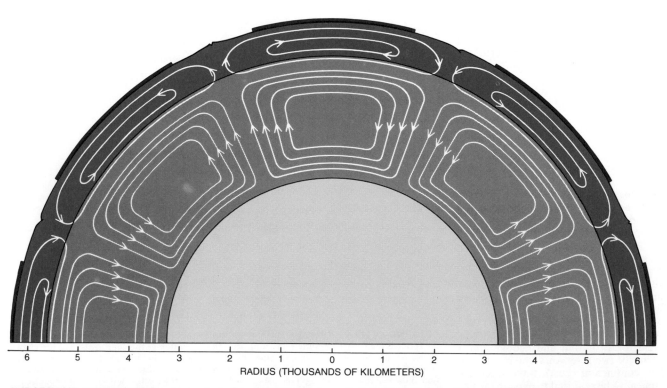

6 5 4 3 2 1 0 1 2 3 4 5 6

RADIUS (THOUSANDS OF KILOMETERS)

IMPORTANT CONSTRAINT is imposed on the convection that could have occurred in the mantle during most of the earth's history by the finding that no more than half of the mantle and possibly as little as a third can be as depleted in large-ion lithophile elements as the basalts erupted at mid-ocean ridges are. Assuming that the entire mantle is in some kind of convective motion, convective cells cannot operate throughout the full depth of the mantle (*top*); convection in the upper mantle must be decoupled from that in the lower (*bottom*).

material is to evaluate the timing and the rate of supply of materials from the mantle to the continental crust, and further to determine the proportion of mantle involved in the processes that have generated the continental crust. This task essentially involves an attempt to determine the comparative abundances of elements in the crust and some portion of the mantle as a function of time. The problem cannot be tackled solely by investigations of either mantle-derived rocks or the continental crust. Instead observations and deductions made from the sampling of materials from both mantle and continental crust must be made mutually compatible with some hypothesis.

The problem was first taken up seriously by Richard L. Armstrong of the University of British Columbia in 1969. In this early attempt to simulate the dynamic nature of continental evolution the continents were considered to have formed in a single episode some four billion years ago and to have been subsequently recycled and exchanged chemically with the mantle in the course of a large number of discrete events. Armstrong's model was able to reproduce some of the features of the isotopic chemistry of the mantle known at that time, but the model is not consistent with what is now known about the pattern of crustal growth.

Recently we modeled the isotopic evolution of the mantle on the basis of the following three assumptions about continental evolution: (1) There is apparently very little continental crust more than 3.8 billion years old. (2) The continents have grown more or less continuously for the past 3.8 billion years. (3) The growth rate of the continents achieved a maximum between 2.5 and three billion years ago (a deduction that follows from the age-distribution pattern in the continents).

A precise knowledge of the abundances of the large-ion lithophile elements in the continental crust is clearly critical to such models. The abundances of these trace elements (particularly the heat-producing radioactive ones) in the continents, however, are much harder to estimate than might be expected, because of their inhomogeneous distribution in the continents. For example, it has become clear from heat-flow studies and geochemical sampling that potassium, thorium and uranium are much more concentrated in the upper part of the crust than in the lower parts.

Acceptable models must also reproduce the isotopic characteristics of the residual mantle and the continents. Our main concern at first was to reproduce the strontium- and neodymium-isotope characteristics of mid-ocean-ridge basalts, because these must by virtue of their large volume provide the best estimate of the isotopic composition of the suboceanic upper mantle. In addition they are generated from the most depleted part of the mantle currently sampled by volcanism. The models we have investigated have assumed continuous differentiation, with the vigor of material transport in the mantle having declined as a function of time parallel to the decline of heat production in the earth. The delay in the stabilization of continental material until approximately 3.8 billion years ago is presumably a result of highly efficient recycling early in the earth's history.

The most important conclusion to emerge from such a model is that no more than half and possibly as little as a third of the mantle can be as depleted in large-ion lithophile elements as the part that supplies basalts at mid-ocean ridges. This conclusion, which is consistent with the findings of DePaolo and Wasserburg at Cal Tech, imposes a ma-

jor constraint on the nature of the convection that could have occurred in the mantle during a large portion of the earth's history. It is difficult to see how cellular convection operating throughout the mantle could have operated to produce depletion in only a limited portion of it. If the entire mantle is indeed involved in some kind of convective motion, as most geophysicists believe, the convection in one portion of it (presumably the upper mantle) must be efficiently decoupled from the convection in the other (presumably the lower mantle). The portion of mantle that has been involved in the formation of the continents has become generally depleted in large-ion lithophile elements, but it is much more depleted in the heat-producing elements potassium, thorium and uranium than it is in strontium, neodymium and samarium.

This last point can be demonstrated by plotting the percent of large-ion lithophile elements originally in the mantle but now residing in the continents as a function of time [see top illustration on page 117]. About 70 percent of the potassium, thorium and uranium originally present in what is now the residual mantle has been transported to the continents. Thus the intrinsic heat production in this residual mantle has declined faster than if chemical differentiation of the continents had not occurred. In spite of the fact that the residual mantle now has a small intrinsic heat production, it continues to operate as the major heat sink of the earth during the creation and cooling of plates generated along the mid-ocean ridges. The heat dissipated during this process arises in part from the radioactive decay of trace elements in less depleted parts of the earth (perhaps in the lower mantle), augmented by some poorly defined yet significant amount of heat released by the cooling of the entire earth.

10

Yellowstone Park as a Window on the Earth's Interior

by Robert B. Smith and Robert L. Christiansen
February 1980

Yellowstone is a "hot spot" in the earth's crust. Its strong volcanic and tectonic activity makes it a unique location for the study of processes that originate deep in the earth

Yellowstone National Park is part of the most seismically active region of the Rocky Mountains. Covering 8,950 square kilometers of Wyoming, Montana and Idaho, Yellowstone gets its name from the brightly colored products of the alteration of its volcanic rocks by steam and hot water. Most of the world's volcanic and tectonic activity is found near the boundaries of the rigid plates that make up the lithosphere: the solid surface of the earth. Some volcanism is nonetheless found at "hot spots" far from the plate boundaries. Yellowstone, which is 2,000 kilometers from the western boundary of the North American plate, is one of these hot spots. And unlike the hot spot represented by the volcanically active island of Hawaii in the Hawaiian Is-

lands, Yellowstone is surrounded not by water but by land. The distinctive geophysical features of Yellowstone make the park a unique natural laboratory for studying the interior of the earth.

Yellowstone spans the continental divide at the juncture of the physiographically distinct northern and middle Rocky Mountains. Within the park is the Yellowstone Plateau, a forested area of 6,500 square kilometers with an average elevation of 2,000 meters. The plateau was formed out of the accumulation of rhyolite and basalt, common volcanic rocks that differ greatly in composition. Where more than 72 percent of the rhyolite consists of silicon dioxides only 50 percent of the basalt consists of them. The plateau is flanked on the north, east and south by mountains that

rise to 4,000 meters. To the west and the southwest the terrain gradually decreases in elevation through a transitional area named Island Park to merge with an arid region of low relief: the Snake River Plain of southern Idaho. Like Yellowstone, the Snake River Plain is a region of active young volcanism. Unlike Yellowstone, however, the plain consists of much more basalt than rhyolite.

The framework for the volcanic and tectonic evolution of Yellowstone was established in the late Mesozoic era, a period of gross geological deformation in the western U.S. Sections of the earth's crust with a long history of stability were compressed and shortened to form large folds and overthrust faults. The deformation culminated 65 million years ago in the Laramide orogeny, which elevated huge blocks of crust to form the middle and southern Rockies. The crustal blocks were separated from basins that were later filled with thick sediment.

In the aftermath of the orogeny volcanism was common in the northern Rockies until about 40 million years ago. The volcanic outflows consisted chiefly of andesite: a volcanic rock with a silicon dioxide content intermediate between rhyolite and basalt. After a hiatus of almost 40 million years a new period of volcanism began in Yellowstone and Island Park.

Several times in the past two million years magma, or fluid rock, has filled immense chambers under the plateau. The now partially crystallized and solidified magma is the source of heat of the numerous hydrothermal features in Yellowstone National Park: geysers, hot springs, mud pots and fumaroles (steam vents). Over the past two million years thousands of cubic kilometers of rhyolitic magma has erupted to the surface. The average rate of magma production has been comparable to the rate at the most active volcanic regions of the

MAP OF NORTHWESTERN U.S. shows the location of Yellowstone Park in Wyoming, Montana and Idaho. The park is part of the most seismically active region in the Rocky Mountains.

earth, including Hawaii, Iceland and the mid-ocean ridges. The volcanism of Yellowstone is more episodic. Periods of voluminous eruption lasting for only a few hours, days or months are separated by quiescent intervals lasting for as much as hundreds of thousands of years. The predominantly basaltic volcanic activity of the oceanic regions is much more continuous.

A major advance in the understanding of the volcanic evolution of Yellowstone came with the discovery that most of the rhyolite was erupted not as lava flows but as particulate flows of volcanic ash and hot gas. Work by one of us (Christiansen) and H. Richard Blank, Jr., of the U.S. Geological Survey has shown that most of the rhyolite erupted in three catastrophic cycles over the past two million years. In each cycle many ash flows erupted in such a short time that they cooled together as a unit, forming distinctive patterns of welding and crystallizing.

Each cycle began with intermittent lava flows, climaxed in a catastrophic ejection of fragmented material and ended with more lava flows. The voluminous hot-ash flows were on a scale known nowhere else in recorded geological history. The ash, which flowed for tens of kilometers, welded to form hard rhyolites covering thousands of square kilometers. The massive eruptions of each cycle considerably drained the sub-

surface magma chambers, causing the chamber roofs to collapse to form huge calderas: craterlike basins tens of kilometers across. In the course of these explosive eruptions fragments of glassy and crystalline volcanic material were thrown high into the atmosphere and carried for thousands of kilometers. Remnants of these materials have been found as far away as Saskatchewan, Texas and California. Subsequent eruptions of rhyolitic lava have partially filled the calderas.

The ages of the volcanic rocks created in the three cycles of volcanism were determined by John D. Obradovich of the Geological Survey. He dated the units by measuring the concentration of the radioactive isotope of potassium with respect to its decay product argon. This dating in conjunction with geological mapping and stratigraphy suggests that the first and most voluminous cycle of volcanism began about 2.2 million years ago with small eruptions of rhyolite and basalt and reached its climax two million years ago with the first catastrophic ash-flow eruption. The resulting cooling unit, named the Huckleberry Ridge Tuff, has a volume of more than 2,500 cubic kilometers. Such a large volume of igneous material erupting in such a short time establishes the existence of a large magma chamber in the upper crust. The roof collapsed as the chamber ejected magma, although

the resulting caldera, which extended across Island Park and the Yellowstone Plateau, has been largely covered by younger volcanic rock.

The second cycle of volcanism was the least productive of the three. It began with eruptions of rhyolitic lava in an area inside the first caldera in the northern part of Island Park. The climactic ash-flow eruption, which formed the Mesa Falls Tuff, came 1.2 million years ago and gave rise to a cooling unit with a volume of more than 280 cubic kilometers. After the first two cycles of volcanism the center of volcanic activity shifted entirely away from Island Park to the Yellowstone Plateau. Over the past million years the chambers of rhyolitic magma under Island Park solidified. Over the past 200,000 years tectonic fractures have penetrated the old calderas and the underlying solidified magma bodies. As a result basalt has erupted through the fractured caldera floors.

The record of the third volcanic cycle, which began about 1.2 million years ago, is more complete than the record of the other two. For 600,000 years rhyolitic lava erupted on the Yellowstone Plateau intermittently. It poured out of a slowly forming set of ringlike fractures outlining an area that was to collapse subsequently to form the third-cycle caldera. These events suggest that a large magma chamber was forming in

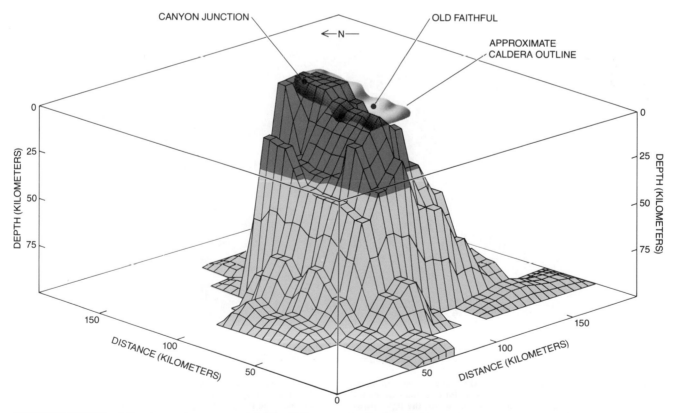

DENSITY MODEL of the crust and upper mantle under Yellowstone shows a 30-kilometer column of material whose peak density is .2 gram per cubic centimeter less than the density of the material that surrounds it laterally. Under the 30-kilometer column is a column of material whose maximum density is .1 gram per cubic centimeter less than the density of the material that surrounds it laterally.

the upper crust. The roof of the chamber domed, stretched and sagged periodically to form the ring fractures. By 600,000 years ago rhyolitic lava had broken through all parts of the ring-fracture system. The conditions were right for a climactic eruption that expelled a great volume of magma through the ring fractures.

What triggers such a large pyroclastic eruption is not completely clear. Magma is a silicate melt containing small amounts of water vapor, other dissolved gases and (if the temperatures are low enough) silicate crystals. In a tumescing system such as the Yellowstone one the ring fractures could propagate downward, eventually penetrating the main magma chamber and thereby reducing the pressure of the magma so that the dissolved gases would be exsolved. Another possibility is that deeply circulating ground water could enter the magma

and saturate it. Either process could initiate a chain reaction of oversaturation, exsolution and degassing. The degassing would push frothing magma up through the ring fractures and reduce the pressure still further.

Whatever the triggering mechanism may actually be, it is clear that once degassing begins it will probably continue at a rate constrained only by the dimensions of the fissures carrying magma to the surface. Magma will cease to erupt only when the pressure in the chamber diminishes to an equilibrium value or when the limits of viscosity allowing the flow of magma are exceeded. In the course of the brief episode of explosive degassing a large amount of magma in the chamber is expanded by frothing and is then chilled and shattered by the explosive release of pressure. The quenched magma is expelled to the surface at high velocities in the form of

glassy and crystalline fragments. Also expelled to the surface is rock torn out of the walls of the venting fissures.

The climactic eruption of the third cycle expelled 1,000 cubic kilometers of magma to form the Lava Creek Tuff. Detailed stratigraphic analysis reveals that the tuff consists of not one but two sheetlike deposits of ash that erupted so close in time that they welded and crystallized as a single cooling unit. Moreover, the caldera that formed out of the collapsed roof of the Lava Creek Tuff magma chamber has two adjacent and almost overlapping zones of ring fractures. The zones correspond to two accumulations at the top of the body of magma that fueled the Lava Creek Tuff eruptions. The rapid drainage of the main chamber caused the roof to collapse along the two overlapping ring-fracture zones, giving rise to a caldera

SATELLITE IMAGE OF YELLOWSTONE and the eastern Snake River Plain at the left is a composite false-color one made in the fall of 1978. The dark reddish brown areas are forests, the light brown areas open fields and the white areas either mountains or hydrothermal features. The map of the same area at the right gives the sites of volcanism, the major faults and the types of volcanic rock. The names of the major fault zones are in red. Yellowstone has had three cycles of volcanism in the past three million years. Each climactic eruption created a caldera, or large craterlike depression, whose boundary is in blue. The blue numbers correspond to the cycle of volcanism in

45 kilometers wide and 75 kilometers long.

If the Yellowstone caldera had remained a basin that later filled with sediment and fresh lava, little about the double-ring structure would be known today. The resurgence of magma in the chamber, however, uplifted the central unfractured part of each ring so that the parts form two domes. The floor of the caldera was uplifted immediately after its collapse. The ash-flow eruption, the collapse of the caldera, the resurgence of magma in the chamber and the doming of the caldera floor all took place within a few thousand years (a span too short to be resolved by the potassium-argon dating method).

Over the past 600,000 years both segments of the caldera filled with sediment and rhyolitic lava that has intermittently erupted. Much of the caldera basin is covered by rhyolitic lava that has

erupted over the past 150,000 years. The source of the eruptions is along two systems of fissures that extend across the caldera from faults outside it. With few exceptions the fissures have vented rhyolitic magma only where they intersect the caldera's ring fractures. This fact suggests that magma at the top of the Yellowstone chamber has been crystallizing and solidifying over the past 150,000 years. As the solidified crust becomes rigid enough to fracture, faults break it, allowing deeper magma to rise to the surface and erupt.

The pattern in which heat flows out of the earth by conduction provides important clues to thermal processes in the earth's crust and the underlying mantle. The rate of the heat flow is measured in milliwatts per square meter. The value of the heat flow characterizes a region's stage of tectonic evolution. For example, tectonically stable parts of the

Rockies near Yellowstone have an average heat flow of about 60 milliwatts per square meter, which is close to the global average. On the other hand, large areas of western North America that have been tectonically active over the past 17 million years have a heat flow of about 100 milliwatts per square meter. Parts of the Snake River Plain have a heat flow of 150 milliwatts per square meter.

Measurements of the conductive heat flow are complicated by the flow of water in the crust. For example, the circulation of ground water in subsurface aquifers can interfere with the measurements by decreasing the heat flow. By the same token the thermal energy transported by convective circulation in hot springs and geysers can disturb the measurements of conductive heat flow. The hot springs and geysers are found in basins that constitute the sur-

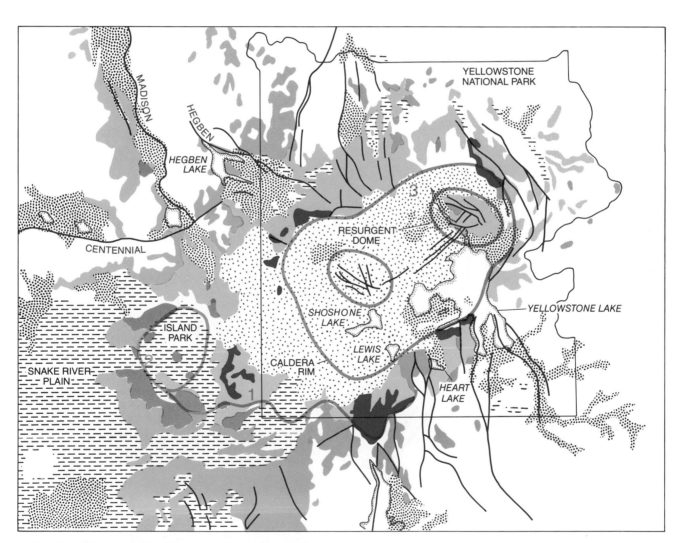

which the caldera formed. The dotted areas are alluvium (sediment deposited by flowing water); the dashed areas are basalt. The green, gray and red areas are accumulations of rhyolite (a glassy volcanic rock similar in composition to granite). The green areas are rhyolite that erupted in the first cycle of volcanic activity, the gray areas rhyo-

lite that erupted in the second cycle and the dark red areas rhyolitic lava that erupted in the third cycle before the caldera formed. The light red areas are rhyolitic ash flows emplaced in the caldera-forming climax of the third cycle of volcanic activity; the red dots are rhyolitic lava erupted in the third cycle after the caldera formed.

face discharge of hot water circulating along systems of deep fractures, including the ring fractures of the caldera and the tectonic fractures that intersect the ring fractures radially.

Such hydrothermal activity transports heat from deep sources not by conduction but by convection. The convective heat flow has been estimated by Donald E. White, R. D. Fournier and Alfred H. Truesdell of the Geological Survey. They measured the amount of water flowing from the area and the concentration in the water of a stable constituent such as chloride. Chloride is abundant in the deep thermal reservoir but scarce in the normal ground water. From measurements and theoretical models of the temperature of liquid water in the underground reservoir a mass unit of chloride in water at depth can be equated with the enthalpy, or heat content, of the liquid. In this way

the convective heat flow out of a hydrothermal basin can be estimated from the total mass discharge of the chloride in the surface drainage of the basin. The heat discharged from all chloride-enriched springs in Yellowstone is about 5.3×10^9 watts. The conclusion is that the average convective heat flow out of the caldera is at least 1,800 milliwatts per square meter, which is more than 20 times the continental average of conductive heat flow.

The southeastern sector of the caldera is flooded by Yellowstone Lake, below which lies an extensive reservoir of hot water. The conductive heat flow out of the lake floor was easy to determine because a 300-meter layer of sediment that is relatively impermeable to the flow of water inhibits convection. The lowest heat flow (120 milliwatts per square meter, or twice the regional average) is at the south end of the lake. Five kilome-

ters inside the boundary of the caldera the heat flow rises to 700 milliwatts per square meter. This dramatic increase is compelling evidence for the lateral extent of the heat source in the shallow crust. Moreover, heat-flow values of some 1,500 milliwatts per square meter at West Thumb and Mary Bay probably reflect the flow of heat out of a shallow hydrothermal system under the lake. The measurement of temperatures of 104 degrees Celsius only four meters below the lake floor, together with the results of seismic profiling (in which the floor is bombarded with sound waves and the resulting echoes are recorded), reveals the possibility of steam-filled sediments at the top of a shallow hydrothermal system.

The total heat flow (conductive and convective together) through the Yellowstone caldera is due both to a high regional heat flow and to thermal energy

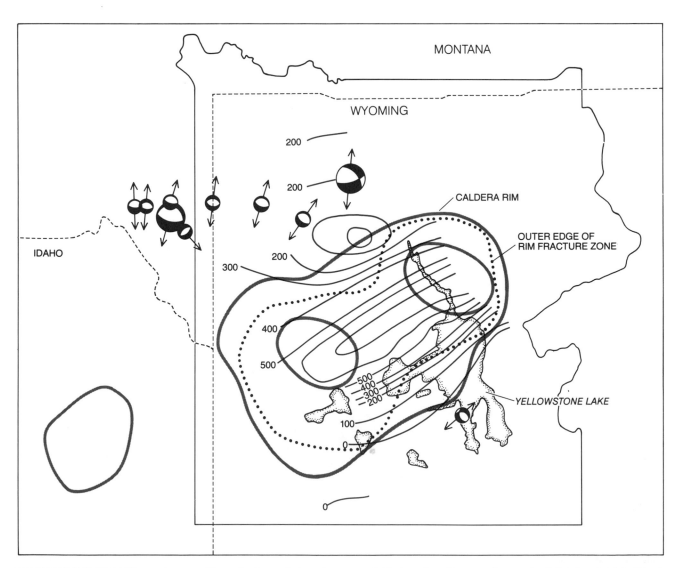

BALLOON DIAGRAMS are stereographic projections of ground motion that help geologists to determine the orientation of faults involved in earthquakes. The size of the balloon gives the strength of the earthquakes at that location. The black area of the balloon indicates where the ground was compressed or became denser; the white **area indicates where the ground was extended or became less dense. The arrow marks the inferred direction of the horizontal component of the ground motion. The colored contour lines give the vertical displacement (in millimeters) of the surface of Yellowstone National Park with respect to a bench mark on the eastern side of the park.**

from shallow localized sources in the crust. The heat transported to the surface by the convective hydrothermal system for at least the past 20,000 years (and probably for much longer) is a result of more than just the cooling of the volcanic rocks of the plateau. Liquid magma clearly existed quite recently in the chamber under Yellowstone and is possibly still there today. The large flow of heat through the caldera is due to the cooling and crystallization of a body of rhyolitic magma at a depth of a few kilometers and to the circulation of ground water deep enough in the caldera fractures to heat and drive the hydrothermal convection.

Earthquakes in Yellowstone are the result of brittle fracturing and faulting in the upper 10 to 20 kilometers of the crust. Investigations of the location, length, displacement and age of the faults in the Yellowstone area and the surrounding region reveal much about their tectonic history. The tectonic activity that began about 17 million years ago in much of the western U.S. has resulted both in the extension of the crust and in significant uplift. For example, since that time the Great Basin of Utah and Nevada has been elevated about 1.5 kilometers.

The uplift and the concomitant extension sustain the topography of parallel valleys and mountain ranges that are bounded by large faults such as the Wasatch Front of Utah and the Sierran Front of California and Nevada. Yellowstone, which lies in the northeastern part of the uplifted expanse, has undergone similar topographical changes. The spectacular escarpment on the east side of the Teton Range reflects a fault zone that extends to the north below the rhyolitic flows of the Yellowstone caldera. Over a period of at least 10 million years the rocks on one side of this fault have been vertically displaced by more than four kilometers with respect to the rocks on the other side. North and west of the caldera the trends of the faults shift to northwesterly ones. And farther west the Hebgen Lake and Centennial faults run westward at 90 degrees to the Teton fault. Internal structures of the Centennial and Teton ranges tilt toward the axis of the Snake River Plain and are buried by young volcanic rocks near the plain's margins. That suggests a crustal downwarping.

In other words, Yellowstone constitutes the intersection of three tectonic trends: the northeast-trending downwarp of the Snake River Plain, the north-trending faults south of the Yellowstone Plateau and the east-to-southeast-trending faults to the west and the north. The intersection of the trends

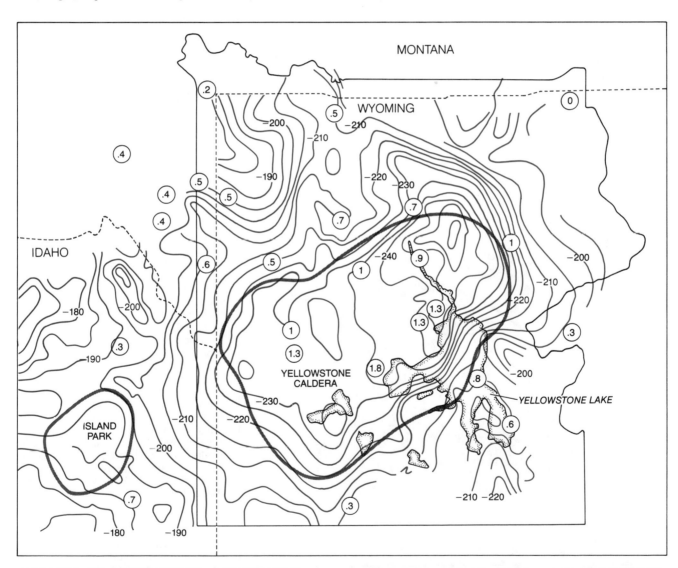

CONTOURS OF GRAVITATIONAL ACCELERATION measured in milligals (a milligal is equal to 10^{-3} centimeter per second squared) reveal that the material under the caldera is much less dense than the surrounding material. Each circled number gives the delay in the travel time of an earthquake compression wave (*P* wave) recorded on a seismograph at the circle. In general a delayed signal corresponds to a decrease in velocity of the material under the seismograph. The substantial delays of as much as 1.8 seconds (from earthquakes between 2,000 and 10,500 kilometers away) at seismographs on the caldera suggest the presence of low-velocity material under it.

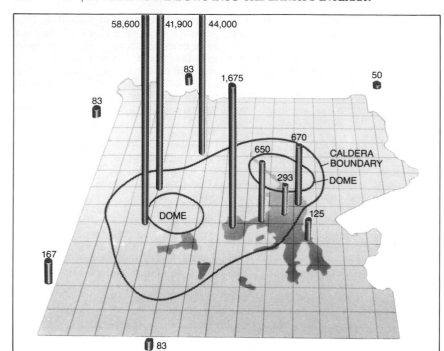

FLOW OF HEAT out of the Yellowstone Plateau, resulting from conduction and convection, is given in milliwatts per square meter. The heat flow out of the bottom of Yellowstone Lake, which was measured by Paul Morgan and D. D. Blackwell of Southern Methodist University, is due only to conduction because impermeable sediments inhibit hydrothermal convection.

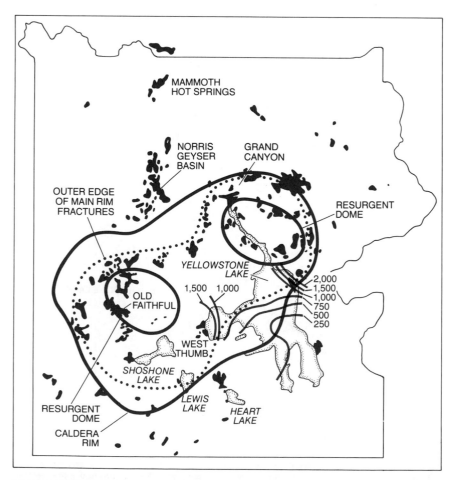

MAP OF HYDROTHERMAL FEATURES of Yellowstone (geysers, steam vents and hot springs) shows the heat-flow contours (*color*) in milliwatts per square meter for Yellowstone Lake. The caldera that formed in the third cycle of volcanism has two ringlike fracture zones.

suggests that Yellowstone lies at the focus of fracturing and stress that helped to localize much of the volcanism and crustal deformation. Extensive earthquake activity indicates that many of the faults are still active.

The exact locations of older earthquakes in Yellowstone National Park are not known because there were no seismographs in the park before the early 1960's, but the general distribution of the earthquake epicenters can be reconstructed from personal accounts of the intensity of the tremors. Such reports indicate that in the past century there have been many more earthquakes in and near Yellowstone than there have been in the surrounding areas of Montana, Wyoming and Idaho. The strongest recorded earthquake in the Rocky Mountains is the Hebgen Lake shock of 1959, the epicenter of which was a few kilometers west of Yellowstone. With a magnitude of 7.1 on the Richter scale, the earthquake was felt over an area of 1.5 million square kilometers. The surface was displaced by as much as six meters along one of two large faults associated with earthquake. The shock waves triggered a massive avalanche of rocks that dammed the Madison River; they also tilted the ground along the northern shore of Hebgen Lake downward by as much as six meters. Earthquakes as strong as this one have undoubtedly occurred in the area at least once every few hundred or few thousand years.

Some of the earthquakes recorded in the western part of the park over the past 20 years are aftershocks of the 1959 Hebgen Lake shock. Seismic studies, however, reveal a high incidence of earthquakes in a zone extending eastward 50 kilometers from the fault zone of Hebgen Lake to the northwest side of the Yellowstone caldera. The foci of these earthquakes are as deep as 16 kilometers. This zone of seismic activity extends eastward to the vicinity of the Norris Geyser Basin, where the strongest shock had reached a magnitude of 6.

In and near the caldera itself earthquake activity is episodic and clustered: a host of earthquakes occur in a short period of time in a limited area without a main shock. Laboratory models suggest that these "swarms" of earthquakes, which are common in areas of active volcanism, are related to concentrated stresses. In Yellowstone's hydrothermal areas such stresses could be the result of high fluid pressures and high temperatures. The Norris Geyser Basin probably has Yellowstone's shallowest thermal reservoir with the highest temperature (more than 275 degrees C.). The close association of seismic activity and intense hydrothermal activity in the basin suggests that fractures caused by earthquakes may serve as conduits for the flow of hot water above a shallow heat source. Southeast of the Norris Geyser Basin the earthquake activity becomes more scattered, with the maximum depth of the earthquake foci seldom exceeding eight kilometers.

The uplift and subsidence of the crust in the Yellowstone area is the result of tectonic and seismic activity. Absolute values for the uplift and the subsidence cannot be determined because there are no stationary points of reference against which the changes in elevation could be measured. What is known is the change in elevation of areas of the park in relation to certain bench marks established in 1923 and resurveyed in the 1970's. John R. Pelton of the University of Utah and one of us (Smith) found that the contours of changes in elevation outlined an elongated area of 3,500 square kilometers coincident with the Yellowstone caldera. In relation to a bench mark east of the caldera the area has risen as much as 700 millimeters. The maximum measured uplift is between the two resurgent domes, where the peak rate of uplift has been more than 14 millimeters per year. Such a high rate of uplift is comparable to the rates measured for active volcanoes on Hawaii and Iceland.

The uplift could be the result of an increase in the pressure in a confined magma body, the increase in turn being the result of an exsolution of gas or an influx of magma from under the body.

Alternatively, the uplift could be the result of tectonic stresses in the crust that are not specifically related to the movement of magma.

Surveys of the region surrounding Yellowstone reveal that the park itself is part of a larger area that is being uplifted about six millimeters per year, and that the Snake River Plain is subsiding relatively from two to six millimeters per year. In 1977 investigators at the University of Utah began monitoring uplift and subsidence in Yellowstone by measuring the acceleration of gravity at various points. If the density or the mass of the crust does not vary, changes in the acceleration of gravity at the surface can be attributed to changes in elevation. Measurements of gravitational acceleration can be made to a precision of about 3×10^{-3} milligal (a milligal being 10^{-3} centimeter per second squared), which corresponds to a change in elevation of about 10 millimeters. In five or 10 years the investigators will measure the gravitational acceleration at the same points so that the change in elevation can be calculated.

The subsurface geological structure of Yellowstone has been investigated by deep-sounding geophysical techniques that measure variations in density, in the velocity of seismic waves, in electrical conductivity and in magnetic susceptibility (the capacity of a material to become magnetized). Such measurements in conjunction with the record of vol-

canism provide a picture of the physical properties of the crust that influence phenomena at the surface. H. M. Iyer of the Geological Survey has measured the time it takes a distant earthquake compressional wave (a P wave) to travel from the epicenter to seismographs in Yellowstone. If the earth were homogeneous, the travel times would be identical for equivalent distances. Any delay or advance in the travel time indicates a change in the velocity below the surface. Seismographs in the vicinity of the Yellowstone caldera have recorded travel-time delays of as much as 1.8 seconds in P waves from earthquakes between 2,000 and 10,500 kilometers away. Delays of no more than .5 second were recorded in the area surrounding the caldera. Modeling of the P-wave delays under Yellowstone indicates a 15 percent reduction in velocity within the crust and a 5 percent reduction in velocity in a region that may extend as much as 250 kilometers into the upper mantle.

Measurements at 900 stations on the caldera and its immediate surroundings indicate that the gravitational acceleration is 60 milligals less than it would be if the earth below the caldera were homogeneous. Quantitative modeling has shown that the mass deficiency responsible for the 60-milligal anomaly could be due not only to low-density sediments in the caldera but also to low-density material in the subvolcanic basement. Such material could be magma or a solidified low-density igneous body.

Which of these possibilities is the most likely? A mathematical model incorporating both the gravity data and the P-wave time-delay data, employed by Jeffrey A. Evoy of the University of Utah, indicates that Yellowstone is underlain to a depth of at least 250 kilometers by a low-density, low-velocity body. Below this body is denser material. Such a structure is consistent with geological interpretations suggesting that a voluminous shallow chamber may consist at least partly of magma containing much silicon dioxide, whereas smaller pockets of magma with less silicon dioxide in the lower crust and the upper mantle are more dispersed.

Information about temperatures below the surface can be inferred from the magnetic properties of rocks. Below a depth called the Curie depth the rocks have reached a threshold temperature (the Curie temperature) where they are not magnetic. For pure magnetite the Curie temperature is 580 degrees C., and for other common crustal materials it is about 560 degrees. In the Yellowstone area the Curie depth is thought to be 10 kilometers below the surface, but in places under the caldera it may be as shallow as six kilometers. The Curie depth for some other areas of the continental U.S. is between 15 and 30 kilome-

ESCARPMENT OF THE RED CANYON FAULT, which is three kilometers north of Hebgen Lake in Montana and 10 kilometers west of the park, was vertically displaced by several meters by the Hebgen Lake earthquake of August, 1959. The quake had a magnitude of 7.1.

ters. This of course means that the rock under Yellowstone reaches a temperature of 560 degrees at a shallower depth than the rock under these other areas does. Measurements of the electric conductivity of the material under Yellowstone indicate that the material at this depth is a good conductor. Materials at high temperatures often are good conductors, and so the magnetic and electrical investigations are in agreement.

All the geophysical data point to an anomalous crustal and upper-mantle structure under Yellowstone. Low seismic-wave velocities, low densities, high electric conductivity and a shallow Curie depth all favor a crustal body at high temperatures. At between five and 10 kilometers below the surface the crustal material apparently becomes so hot and weak that it cannot fracture, although it can deform by flow and creep. Although it is not known with certainty whether some of this material is currently a liquid, much evidence points to that conclusion.

It is time to relate the geophysical picture of Yellowstone to the tectonic and volcanic history of the surrounding areas, particularly the eastern part of the Snake River Plain. We mentioned above that the Laramide mountain building took place in an environment of crustal shortening with the convergence of two lithospheric plates. The plates were the one forming the floor of the Pacific and the one carrying the continent of North America. At the time of the plates' convergence andesitic volcanism was prevalent on the continent. Over the past 30 million years the large-scale tectonics of the western U.S. has changed substantially because of a gradual transition from the comparatively simple convergence of two plates to the complex interaction of three plates: the North American plate, the Pacific plate and the Juan de Fuca plate. Displacement along the San Andreas fault in California indicates that currently the relative motion between the North American and the Pacific plate is chiefly horizontal. Much of the western U.S. is now undergoing tectonic extension and has experienced both rhyolitic and basaltic volcanism of the kind that has occurred in Yellowstone.

During this period of regional tectonic extension the eastern Snake River Plain has evolved in a remarkable way. Some 15 million years ago the eastern Snake River Plain and the Yellowstone Plateau did not exist. Since that time basalt and rhyolite have erupted in a systematic sequence propagating northeastward from near the borders of Idaho, Nevada and Oregon along a line that is now the axis of the eastern Snake River Plain. If an observer could go back to any given time in this period, he would see a "Yellowstone": a topographically high area of voluminous rhyolitic volcanism and extensive thermal and seismic activity. Yet on a subsequent visit he would find "Yellowstone" farther northeast than it had been. As the zone of voluminous rhyolitic volcanism propagates northeastward at the rate of between two and four centimeters per year basaltic volcanism continues to occur periodically in its wake. The regions of past active rhyolitic volcanism, which have subsided, have been flooded by basalts to form the eastern Snake River Plain.

A close association of rhyolitic and basaltic magma has characterized the development of Yellowstone and the Snake River Plain. Many explanations have been put forward to explain the origin of the two kinds of magma, the most promising hypothesis being that two different source materials melted partially to form two distinct liquids, one rhyolitic and the other basaltic.

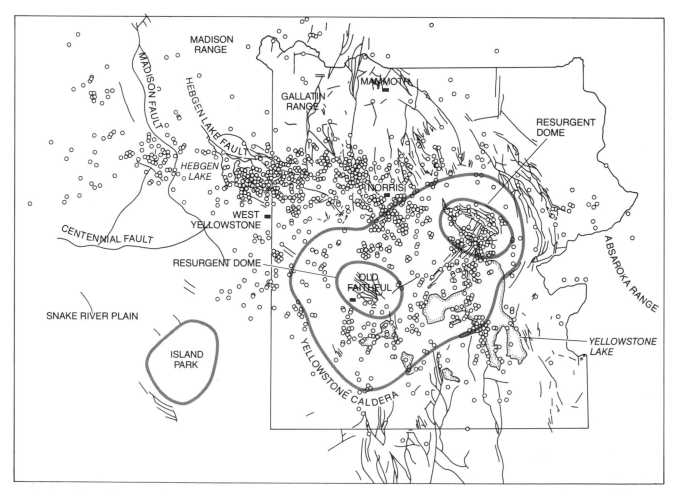

MAP OF EPICENTERS (*black circles*) of more than 1,500 earthquakes with magnitudes on the Richter scale of between 1 and 6 comes from measurements made on seismographs of the U.S. Geological Survey and the University of Utah. The peak focal depth was 16 kilometers.

(This explanation may seem trivial, but it is not. The two kinds of magma could conceivably have formed not out of two materials but out of a single material with two immiscible ingredients.) The basaltic magma is quite similar to the basaltic magma that forms under oceanic islands; in both settings the basaltic magmas are partial melts of the earth's upper mantle. The origin of the rhyolitic magma is probably the partial melting of metamorphic rocks of the earth's lower crust.

The generation of basaltic magma by the partial melting of the upper mantle is the fundamental process that drives the Yellowstone magmatic system. The continuing uplift and tectonic extension of broad areas of the Yellowstone region requires the upward displacement of the mantle. As the crust is stretched and the underlying mantle moves upward the reduction of pressure may cause materials with low melting points to begin melting. Basalt is the result of

this partial melting, as it is in the floor of the deep ocean, where the mantle lies only a few kilometers below the solid surface. The flowing of the basaltic magma from the mantle to the crust transports heat upward. The material of the lower crust is under less pressure and has a higher concentration of silicon dioxides than the underlying mantle. As a result when this material melts, it yields magma with a higher concentration of silicon dioxides. Such magma has a lower temperature, a lower density and a greater viscosity than basaltic magma, and so its buoyancy causes it to rise and accumulate as a large mass in the upper crust.

It seems that rhyolitic magma chambers, such as those that have underlain Yellowstone and Island Park for the past two million years, must have evolved in each successive area of rhyolitic volcanism along the propagating axis of the eastern Snake River Plain. This kind of volcanic propagation has

also occurred on Hawaii. The volcanism of Yellowstone and that of Hawaii are remarkably similar, with the difference that the Yellowstone system has evolved not under oceanic crust but under continental crust.

In both Yellowstone and Hawaii an area of concentrated melting seems to be fixed and localized within the earth under a moving lithospheric plate. The movement of a plate over each of these hot spots has left a trail of extinct volcanoes, which are milestones marking the passage of the plate. All the islands of the Hawaiian chain were created by a single source of magma over which the Pacific plate passed as it proceeded on a course roughly toward the northwest. The relative motion of the tectonic plates has been reconstructed in detail, but the motion of one plate with respect to another cannot readily be translated into motion with respect to the earth's interior, except in relation to the hot spots anchored in the mantle.

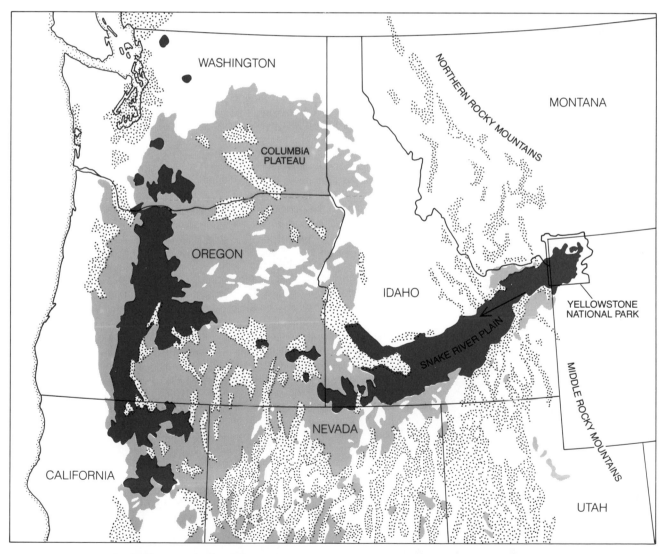

MAP OF THE YELLOWSTONE AREA relates the young tectonic and volcanic features of the park to the tectonic and volcanic features of the rest of the northwestern U.S. The valleys filled with young alluvium (*dots*) reveal mountain ranges and basins bounded by linear faults. The youngest volcanic rocks (younger than two million years) are in color; the older Cenozoic volcanic rocks are in gray. The vector with a magnitude of 4.5 centimeters per year marks the motion of the North American plate with respect to a point deep in the mantle.

The calculated motion of the North American plate with respect to the Hawaiian hot spot is 4.5 centimeters per year to the southwest. The rhyolitic volcanism along the axis of the Snake River Plain and Yellowstone has propagated at about the same rate in the opposite direction.

What is the mechanism that generates hot spots? Several mechanisms have been proposed, but it is difficult to choose among them because the circulation of the material in the mantle is still poorly understood. According to one hypothesis, the hot spots are surface manifestations of "plumes": rising columnar currents of hot material. Another hypothesis considers the hot spots to be the result of a localized overabundance of heat-generating radioactive material in the mantle. According to another explanation, the frictional heating of the base of the lithosphere as it moves over the underlying asthenosphere could generate a hot spot if the conditions were right to sustain the heat. Still another explanation locates the hot spots at the tips of linearly propagating fractures of the lithosphere that are driven by extensional forces. The hot spot anchored under Yellowstone is more accessible to investigation than most of the other continental hot spots, and so the continuing studies of the park should help in identifying the generative mechanism.

Does the picture of Yellowstone that has emerged out of investigations of its volcanic and tectonic history suggest anything about the future evolution of the park? We think it does, although not enough information has yet been gathered to make detailed predictions. The extent of past volcanism and uplift, the length of time between volcanic eruptions, the probable existence of a strong heat source just below the surface, the swarms of earthquakes, the rhyolitic volcanism in the caldera and the current rapid rate of uplift all point to future volcanic activity in the Yellowstone Plateau. The high heat flow, the delays in earthquake P waves and other geophysical data are consistent with the probable existence of magma in the crustal chamber under the plateau. The evidence does not indicate what form the future volcanism might take; it could consist of relatively small eruptions of rhyolite and basalt at the margins of the plateau, medium-sized eruptions of rhyolitic lava within the caldera or a major ash flow. Any renewal of volcanism in the park would probably be preceded by phenomena such as localized earthquake activity and increased gas emissions. Future earthquake activity is a certainty, presumably including shocks at least as strong as the Hebgen Lake earthquake. It is also known that the surface of the Yellowstone caldera will continue to undergo deformation.

The Authors

M. NAFI TOKSÖZ ("The Subduction of the Lithosphere") is professor of geophysics at the Massachusetts Institute of Technology, where he heads the George R. Wallace, Jr., Geophysical Observatory. Born in Turkey, Toksöz completed his undergraduate studies at the Colorado School of Mines and went on to obtain a Ph.D. in geophysics and electrical engineering from the California Institute of Technology in 1963. He did research at Cal Tech for several years before moving to M.I.T. in 1965. A specialist in seismology, plate tectonics and the structure and evolution of planetary interiors, he has devoted his recent efforts to "extensive theoretical calculations" in an effort to show that "processes involved in the earth's lithosphere and asthenosphere can indeed produce geological and geophysical impressions at the surface."

KEN C. MACDONALD and BRUCE P. LUYENDYK ("The Crest of the East Pacific Rise") are at the University of California at Santa Barbara; Macdonald is associate professor of marine geophysics and Luyendyk is professor of marine geophysics. Macdonald, who says he was "lured into oceanography by a love for diving, travel and quantitative natural sciences," was graduated from the University of California at Berkeley with a degree in engineering and received his Ph.D. (in oceanography) jointly from the Massachusetts Institute of Technology and the Woods Hole Oceanographic Institution. He worked at the Scripps Institution of Oceanography from 1975 to 1979 and still has a joint research appointment there. Luyendyk received his bachelor's degree in geology from San Diego State University and his Ph.D. in oceanography from the Scripps Institution of Oceanography. He worked at Woods Hole from 1969 to 1973 and then went to Santa Barbara, where, he writes, "I have made a partial transition into becoming more involved in on-land geology and geophysics." His projects in that area include studying "the paleomagnetism and tectonics in southern California and the magnetic properties of ophiolites (on-land exposures of ocean crust)."

KEVIN C. BURKE and J. TUZO WILSON ("Hot Spots on the Earth's Surface") are geophysicists with a common interest in plate tectonics. Burke, who is chairman of the department of geological sciences at the State University of New York at Albany, obtained his Ph.D. from University College London in 1953. He has done geological fieldwork in 27 countries. In the 1950's he taught at the University of Ghana and later did research for the British Geological Survey on nuclear raw materials. From 1961 to 1965 he was involved in establishing a geology department at the Univeristy of the West Indies in Jamaica, and from 1965 to 1971 he taught at the University of Ibadan in Nigeria. Wilson, who played a key role in the revival of the theory of continental drift first put forward by Alfred L. Wegener in 1912, is currently director general of the Ontario Science Centre. His degrees include a B.A. from the University of Toronto, a B.A., an M.A., and a D.Sc. from the University of Cambridge and a Ph.D. from Princeton University. He received the Vetlesen Award from Columbia University. Before taking up his present post he was for many years professor of geophysics at Toronto, where he also served as director of the Institute of Earth Sciences. He is currently President of the American Geophysical Union.

DALLAS L. PECK, THOMAS L. WRIGHT and ROBERT W. DECKER ("The Lava Lakes of Kilauea") began their studies of molten "lakes" of lava in 1963, when they were stationed at the Hawaiian Volcano Observatory. Peck is Director of the U.S. Geological Survey. He obtained his bachelor's and master's degrees at the California Institute of Technology and his Ph.D. from Harvard University in 1960. Peck then did studies for the Geological Survey in western Oregon, in the Sierra Nevada granite-batholith area of California and in Hawaii. Wright is a staff geologist with the Survey. He did his undergraduate work at Pomona College and received his Ph.D. from Johns Hopkins University in 1961. He then did studies for the Survey in the Cascade Range of Washington and in Hawaii. Since leaving

Hawaii in 1969 Wright has investigated basalts from the moon, the mid-Atlantic Ridge and the Columbia River. Decker is scientist-in-charge of the U.S. Geological Survey's Hawaiian Volcano Observatory. He did his undergraduate and master's work at the Massachusetts Institute of Technology and obtained his D.Sc. from the Colorado School of Mines in 1953. After teaching and doing volcanological research in Indonesia he joined the faculty of Dartmouth College. While he was at Dartmouth he did field work in Hawaii, Iceland and Central America.

LAURENCE R. KITTLEMAN ("Tephra") is a practicing geologist in Eugene, Ore., and adjunct associate professor of geology at Southern Oregon State College in Ashland. He received his B.S. at Colorado College, his M.S. at the University of Colorado and his Ph.D. from the University of Oregon in 1962, all in geology. He worked for a time as a geologist for the U.S. Atomic Energy Commission, and in 1962 he joined the faculty of the Museum of Natural History at the University of Oregon, serving as curator of geology until 1977 and as director from 1974 to 1977. He has done research on Upper Cenozoic volcanic rocks in Oregon and on the stratigraphy and petrology of volcanic rocks. Kittleman is also interested in the geology of prehistoric human habitations.

ROBERT DECKER and BARBARA DECKER ("The Eruptions of Mount St. Helens") have collaborated on several books and articles about volcanoes. They were both engaged in on-site studies of Mount St. Helens in the spring and summer of 1980; during part of that time Robert Decker was the acting scientist in charge of the U.S. Geological Survey's Mount St. Helens Project. He studied geology and geophysics at the Massachusetts Institute of Technology and the Colorado School of Mines, obtaining his D.Sc. from the latter institution. He is a past president of the International Association of Volcanology and Chemistry of the Earth's Interior. Before he joined the Geological Survey in 1979 he was professor of geophysics at Dartmouth College. Barbara Decker, who has a B.A. in journalism from the University of California at Berkeley, is a professional writer. The Deckers (husband and wife) now live in Hawaii, where he heads the Geological Survey's Hawaiian Volcano Observatory.

PETER J. WYLLIE ("The Earth's Mantle") is Homer J. Livingston Professor and Chairman of the Department of Geophysical Sciences at the University of Chicago. Born in London, Wyllie studied at the University of St. Andrews in Scotland, receiving his Ph.D. in geology there in 1958. He was originally atracted to geology, he writes, "because I wanted to see the world and work outdoors. At first this worked out well. I spent two years driving a dog sledge in geological survey work on an expedition to Greenland. Then I became involved in experimental geology, synthesizing rocks and repro-

ducing deep-seated rock processes in high-pressure/high-temperature apparatus. Since then my work has been mainly in the laboratory, or in front of a typewriter trying to catch up with a backlog of research papers arising from the data supplied by the apparatus." Before moving to Chicago in 1965, Wyllie taught and did research at the University of Leeds and Pennsylvania State University.

KEITH G. COX ("Kimberlite Pipes") is university lecturer in the department of geology and mineralogy at the University of Oxford. He studied for his bachelor's degree at Oxford, and then in 1956 became a research student in the Research Institute of African Geology at the University of Leeds. An interest in igneous geology led him to do fieldwork in the Karroo volcanic province of South Africa. He writes: "During the course of these studies I visited Lesotho to see the Karroo basalts there and, quite by accident, was directed by a local diamond prospector to the Matsoku kimberlite pipe, at that time unknown to science. I then took up the study of kimberlites as a scientific 'hobby' to accompany my main interest in basaltic rocks, and in 1963 I led an expedition to Matsoku." After leaving Leeds, Cox spent seven years lecturing at the University of Edinburgh before returning to Oxford. He is the coauthor of two textbooks on igneous petrology, one introductory and one advanced, and is an editor of *Journal of Petrology* and of *Earth and Planetary Science Letters*.

R. K. O'NIONS, P. J. HAMILTON, and NORMAN M. EVENSEN ("The Chemical Evolution of the Earth's Mantle") established their working association at the Lamont-Doherty Geological Observatory of Columbia University. O'Nions and Hamilton are now at the Univeristy of Cambridge, where O'Nions is Royal Society Research Professor and Hamilton is research associate in the department of earth sciences; Evensen is assistant professor of geology at the University of Toronto. O'Nions did his undergraduate work at the Univeristy of Nottingham and obtained his Ph.D. (in geochemistry) from the University of Alberta in 1969. He taught at the University of Oxford until 1975, when he began a four-year association with Lamont-Doherty. Hamilton was graduated from King's College at the University of London in 1972 and received his D.Phil. from Oxford in 1975. He was at Lamont-Doherty from then until this year. Evensen's degrees are from the University of Minnesota. He was at Lamont-Doherty from 1974 until this year.

ROBERT B. SMITH and ROBERT L. CHRISTIANSEN ("Yellowstone Park as a Window on the Earth's Interior") share a professional interest in the Yellowstone area. Smith is professor of geophysics and director of seismograph stations at the Univeristy of Utah. He received his Ph.D. in 1967 from the University of Utah. Smith was a visiting scientist at the Lamont-Doherty Geological Observatory of Columbia

University in 1967 and a visiting professor at the Swiss Federal Institute of Technology-Zurich in 1967 and 1977. His interests are in earthquakes, seismic structure of the crust and mantle, and mechanics of volcanism and mountain building. Christiansen has been associated with the U.S. Geological Survey since 1961, when he obtained his Ph.D. in geology from Stanford University. From 1976 through 1979 he managed the Geological Survey's Geothermal Research Program.

Beginning with the first eruptions of Mount St. Helens in March of 1980, he coordinated the establishment and development of monitoring and scientific investigations of that volcano through the first several months of its activity. A specialist in volcanology and igneous petrology, he has now returned to full-time research. Smith and Christiansen would like to acknowledge the supportive encouragement of their work by the National Park Service.

BIBLIOGRAPHIES

I VOLCANOES AND PLATE TECTONICS

1. The Subduction of the Lithosphere

SEISMOLOGY AND THE NEW GLOBAL TECTONICS. Bryan Isacks, Jack Oliver and Lynn R. Sykes in *Journal of Geophysical Research*, Vol. 73, No. 18, pages 5855–5899; September 15, 1968.

MOUNTAIN BELTS AND THE NEW GLOBAL TECTONICS. John F. Dewey and John M. Bird in *Journal of Geophysical Research*, Vol. 75, No. 14, pages 2625–2647; May 10, 1970.

TEMPERATURE FIELD AND GEOPHYSICAL EFFECTS OF A DOWNGOING SLAB. M. Nafi Toksöz, John W. Minear and Bruce R. Julian in *Journal of Geophysical Research*, Vol. 76, No. 5, pages 1113–1138; February 10, 1971.

EVOLUTION OF THE DOWNGOING LITHOSPHERE AND THE MECHANISMS OF DEEP FOCUS EARTHQUAKES. M. Nafi Toksöz, Norman H. Sleep and Albert T. Smith in *The Geophysical Journal of the Royal Astronomical Society*, Vol. 35, Nos. 1–3, pages 285–310; December, 1973.

THE MOVEMENT OF CONTINENTS. Z. Ben-Avraham in *American Scientist*, pages 291–299, May–June, 1981.

INITIATION OF INTRACONTINENTAL SUBDUCTION IN THE HIMALAYA. P. Bird in *Journal of Geophysical Research*, Vol. 83, pages 4975–4987; 1978.

MECHANICS OF SUBDUCTED LITHOSPHERE. G. F. Davies in *Journal of Geophysical Research*, Vol. 85, pages 6304–6318; 1980.

GEOMETRY OF BENIOFF ZONES: LATERAL SEGMENTATION AND DOWNWARDS BENDING OF THE SUBDUCTED LITHOSPHERE. B. L. Isacks and M. Barazangi in *In Island Arcs, Deep Sea Trenches and Back Arc Basins*, edited by M. Talwani and W. C. Pitman III, American Geophysical Union, pages 99–114; 1977.

ON THE FORCES DRIVING PLATE TECTONICS: INFERENCES FROM ABSOLUTE PLATE VELOCITIES AND INTRAPLATE STRESS. S. C. Solomon, N. H. Sleep and R. M. Richardson in *Geophysical Journal of the Royal Astronomical Society*, Vol. 42, pages 769–801; 1975.

OCEANIC RIDGES AND ARCS, GEODYNAMIC PROCESSES. Edited by M. N. Toksöz, S. Uyeda and J. Francheteau. Elsevier Scientific Publishing Company, 1980.

2. The Crest of the East Pacific Rise

MASSIVE DEEP-SEA SULPHIDE ORE DEPOSITS DISCOVERED ON THE EAST PACIFIC RISE. J. Francheteau, H. D. Needham, P. Choukroune, T. Juteau, M. Séguret, R. D. Ballard, P. J. Fox, W. Normark, A. Carranza, D. Cordoba, J. Guerrero, C. Rangin, H. Bougault, P. Cambon and R. Hekinian in *Nature*, Vol. 277, No. 5697, pages 523–528; February 15, 1979.

SUBMARINE THERMAL SPRINGS ON THE GALÁPAGOS RIFT. John B. Corliss, Jack Dymond, Louis I. Gordon, John M. Edmond, Richard P. von Herzen, Robert D. Ballard, Kenneth Green, David Williams, Arnold Bainbridge, Kathy Crane and Tjeerd van Andel in *Science*, Vol. 203, No. 4385, pages 1073–1083; March 16, 1979.

EAST PACIFIC RISE: HOT SPRINGS AND GEOPHYSICAL EXPERIMENTS. The RISE Project Group: F. N. Spiess, Ken C. Macdonald, T. Atwater, R. Ballard, A. Carranza, D. Cordoba, C. Cox, V. M. Diaz Garcia, J. Francheteau, J. Guerrero, J. Hawkins, R. Haymon, R. Hessler, T. Juteau, M. Kastner, R. Larson, B. Luyendyk, J. Macdougall, S. Miller, W. Normark, J. Orcutt and C. Rangin in *Science*, Vol. 207, No. 4438, pages 1421–1433; March 28, 1980.

3. Hot Spots on the Earth's Surface

PHYSICS AND GEOLOGY. J. A. Jacobs, R. D. Russell and J. Tuzo Wilson. McGraw-Hill Book Company, Inc., 1959.

IS THE AFRICAN PLATE STATIONARY? K. Burke and J. T. Wilson in *Nature*, Vol. 239, No. 5372, pages 387–390; October 13, 1972.

TWO TYPES OF MOUNTAIN. J. Tuzo Wilson and Kevin Burke in *Nature*, Vol. 239, No. 5373, pages 448–449; October 20, 1972.

PLUME GENERATED TRIPLE JUNCTIONS: KEY INDICATORS IN APPLYING PLATE TECTONICS TO OLD ROCKS. Kevin Burke and J. F. Dewey in *The Journal of Geology*, Vol. 81, No. 4, pages 406–433; July, 1973.

RELATIVE AND LATITUDINAL MOTION OF ATLANTIC HOT SPOTS. Kevin Burke, W. S. F. Kidd and J. Tuzo Wilson in *Nature*, Vol. 245, No. 5421, pages 133–137; September 21, 1973.

II VOLCANIC PRODUCTS: LAVA, ASH, AND BOMBS

4. The Lava Lakes of Kilauea

PETROLOGY OF THE KILAUEA IKI LAVA LAKE, HAWAII. Donald H. Richter and James G. Moore. Geological Survey Professional Paper 537–B, 1966.

THE ERUPTION OF AUGUST 1963 AND THE FORMATION OF ALAE LAVA LAKE, HAWAII. Dallas L. Peck and W. T. Kinoshita. Geological Survey Professional Paper 935–A, 1976.

5. Tephra

HEKLA: A NOTORIOUS VOLCANO. Sigurdur Thórarinsson. Almenna Bókafélagid, 1970.

MINERALOGY, CORRELATION, AND GRAIN-SIZE DISTRIBUTIONS OF MAZAMA TEPHRA AND OTHER POSTGLACIAL PYROCLASTIC LAYERS, PACIFIC NORTHWEST. Laurence R. Kittleman in *Geological Society of America Bulletin*, Vol. 84, No. 9, pages 2957–2980; September, 1973.

WORLD BIBLIOGRAPHY AND INDEX OF QUATERNARY TEPHROCHRONOLOGY. Edited by J. A. Westgate and C. M. Gold. International Union of Quaternary Research, Printing Services Department, University of Alberta, 1974.

GEOLOGICAL HAZARDS. B. A. Bolt, W. L. Horn, G. A. Macdonald and R. F. Scott, Springer-Verlag, 1975.

VOLCANOES OF THE EARTH. Fred M. Bullard. University of Texas Press, 1976.

VOLCANIC ACTIVITY AND HUMAN ECOLOGY. Edited by Payson D. Sheets and Donald K. Grayson. Academic Press, 1979.

6. The Eruptions of Mount St. Helens

POTENTIAL HAZARDS FROM FUTURE ERUPTIONS OF MOUNT ST. HELENS VOLCANO, WASHINGTON. Dwight R. Crandell and Donal R. Mullineaux. U.S. Geological Survey Bulletin 1383–C, 1978.

VOLCANOES. Robert Decker and Barbara Decker. W. H. Freeman and Company, 1981.

VOLCANOLOGY. Howel Williams and A. R. McBirney. Freeman, Cooper and Co., 1979.

FIRE AND ICE: THE CASCADE VOLCANOES. Stephen L. Harris. Pacific Search Press, 1980.

1980 ERUPTIONS OF MOUNT ST. HELENS. Edited by Peter W. Lipman and Donal R. Mullineaux. U.S. Geological Survey Professional Paper No. 1250, 1982.

III VOLCANIC WINDOWS INTO THE EARTH'S INTERIOR

7. The Earth's Mantle

ULTRAMAFIC AND RELATED ROCKS. Edited by Peter J. Wyllie. John Wiley & Sons, Inc., 1967.

THE DYNAMIC EARTH: TEXTBOOK IN GEOSCIENCES. Peter J. Wyllie, John Wiley & Sons, Inc., 1971.

CONTINENTS ADRIFT: READINGS FROM *Scientific American*. Introductions by J. Tuzo Wilson. W. H. Freeman and Company, 1972.

EARTH, 3rd ed. Frank Press and Raymond Siever. W. H. Freeman and Company, 1982.

EARTH, MECHANICAL PROPERTIES OF; EARTH, STRUCTURE AND COMPOSITION OF; EARTH AS A PLANET; EARTHQUAKES IN *The New Encyclopedia Britannica: Macropaedia, Vol. 6*. Encyclopaedia Britannica, Inc., 1974.

PLANET EARTH: READINGS FROM *Scientific American*. Edited by Frank Press and Raymond Siever. W. H. Freeman and Company, 1974.

8. Kimberlite Pipes

LESOTHO KIMBERLITES. Edited by P. H. Nixon. Lesotho Development Corporation, Maseru, 1973.

OCCURRENCE, MINING AND RECOVERY OF DIAMONDS. A. A. Linari-Linholm. De Beers Consolidated Ltd., Kimberley, 1973.

MODEL OF A KIMBERLITE PIPE. J. B. Hawthorne in *Physics and Chemistry of the Earth*, Vol. 9, pages 1–15; 1975.

SUB-CRATONIC CRUST AND UPPER MANTLE MODELS BASED ON XENOLITH SUITES IN KIMBERLITE AND NEPHELINITIC DIATREMES. John B. Dawson in *Journal of the Geological Society, London*, Vol. 134, Part 2, pages 173–184; November, 1977.

9. The Chemical Evolution of the Earth's Mantle

PRINCIPLES OF ISOTOPE GEOLOGY. Gunter Faure. John Wiley & Sons, Inc., 1977.

THE OLDEST ROCKS AND THE GROWTH OF CONTINENTS. Stephen Moorbath in *Scientific American*, Vol. 236, No. 3, pages 92–104; March, 1977.

THE ORIGIN OF THE EARTH AND THE MOON. Alfred E. Ringwood, Springer-Verlag, 1979.

GEOCHEMICAL AND COSMOCHEMICAL APPLICATIONS OF ND-ISOTOPE ANALYSIS. R. K. O'Nions, S. R. Carter, N. M. Evensen and P. J. Hamilton in *Annual Review of Earth and Planetary Sciences*, Vol. 7, pages 11–38; 1979.

GEOCHEMICAL MODELING OF MANTLE DIFFERENTIATION AND CRUSTAL GROWTH. R. K. O'Nions, N. M. Evensen and P. J. Hamilton in *Journal of Geophysical Research*, Vol. 84, No. B11, pages 6091–6101; October 10, 1979.

10. Yellowstone Park as a Window on the Earth's Interior

VOLCANIC STRATIGRAPHY OF THE QUATERNARY RHYOLITE PLATEAU IN YELLOWSTONE NATIONAL PARK. Robert L. Christiansen and H. Richard Blank, Jr. Geological Survey Professional Paper 729–B, Government Printing Office, 1972.

YELLOWSTONE HOT SPOT: NEW MAGNETIC AND SEISMIC EVIDENCE. R. B. Smith, R. T. Shuey, R. C. Friedline, R. M. Otis and L. B. Alley in *Geology*, Vol. 2, No. 9, pages 451–455; September, 1974.

MAGMA BENEATH YELLOWSTONE NATIONAL PARK. Gordon P. Eaton, Robert L. Christiansen, H. M. Iyer, Andrew M. Pitt, Don R. Mabey, H. Richard Blank, Jr., Isidore Zietz and Mark E. Gettings in *Science*, Vol. 188, No. 4190, pages 787–796; May 23, 1975.

YELLOWSTONE HOT SPOT: CONTEMPORARY TECTONICS AND CRUSTAL PROPERTIES FROM EARTHQUAKE AND AEROMAGNETIC DATA. R. B. Smith, R. T. Shuey, J. R. Pelton and J. P. Bailey in *Journal of Geophysical Research*, Vol. 82, No. 26, pages 3665–3676; September 10, 1977.

INDEX